Lectures in Human Physiology

Larry R. McLean, Ph.D.

CreateSpace Independent Publishing Platform
ISBN-13: 978-1983448485
ISBN-10: 1983448486

Table of Contents

Digestion and Metabolism

Neurophysiology

Introduction to Physiology

All the vital mechanisms, however varied they may be, have only one object, that of preserving constant the conditions of life in the internal environment.

– Claude Bernard

Science of Physiology

We begin by considering how *science* is used to investigate body functions. *Anatomy* studies structure. *Physiology* studies function. Living organisms carry out six functions: preserving structure, metabolism, homeostasis, responsiveness, growth and development, and reproduction. The relationships between structure and function are divided into eleven *anatomical systems*. Each system consists of a set of organs and related structures that work together to carry out a specific function of the body. We will take an anatomical systems approach to the basic functions of the body, but first need to master some preliminary topics that come up over and over again in our study. In the first lecture, we learn about the four basic *tissue types*, the four classes of *biomolecules*, and how *negative feedback* maintains a relatively constant internal environment (*homeostasis*).

Scientific Method

Explain the principles of the scientific method

Scientists study the world using the *scientific method*, an on-going process that constantly builds knowledge and refines our understanding of how it all works. The scientific method begins with observing and asking questions about the world. We observe that the chest rises and falls as we breathe. This leads us to wonder how that happens and why it is important. An explanation for the *observations* or tentative answer to the questions is proposed. This is the *hypothesis* to be tested. We might propose that the rising of the chest expands the lungs, which reduces their internal pressure to allow air to enter.

Predictions are made that can be tested by carefully controlled *experiments* and systematic observation. If the hypothesis is true, the volume of the lungs should increase during inspiration. How can this prediction be tested? We might put a person in a tub of water and measure the volume displaced as they breathe in and out. If the level in the tub rises on inspiration, the chest wall has expanded.

As the experiments and observations are carried out, the results are carefully recorded as *data*. Say, we observe that the level of water in the tub rises. We would carefully record how much the level increased and calculate from this how much the volume of the tub increased. We might have the subject breathe more deeply to see if the volume increased even more. Or, we could have them breathe in air from a bag so that the volume inspired can be compared to the volume increase of the bath. Now, an *analysis* of the results is carried out to see how the data compare with the predictions of the hypothesis.

Since we know from anatomy that the lungs are attached to the chest wall, *expansion* of the lungs should swell the chest and *increase* the volume of the tub of water. The volume inspired from a bag should correspond to the volume increase of the tub. This is what is observed.

The data are consistent with our hypothesis, but still not yet a complete explanation of what happens in breathing. Our analysis of the results causes us to ask more questions. For one, how is it that the body causes the chest wall and the attached lungs to expand? Anatomy gives us clues. How might contraction of the diaphragm that forms the lower wall of the chest cavity change thoracic volume? Anatomy suggests that contraction flattens the diaphragm, expanding the thorax and the lungs that are attached to the chest wall.

Physiology uses the scientific method to answer questions about how the body works: observations lead to hypotheses that are tested by experiment and refined to develop explanations

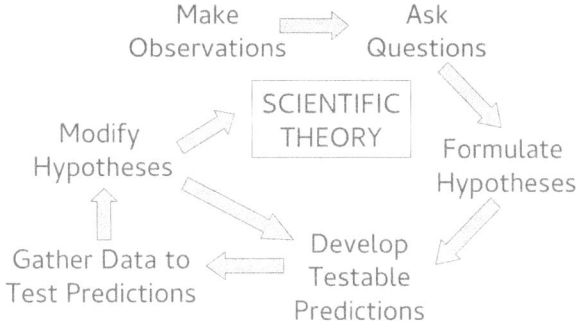

A generally accepted comprehensive explanation of some aspect of nature is a *scientific theory*. Scientific theories are supported by facts that have been repeatedly confirmed by data. They are the foundation of our understanding of nature. For example, atomic theory tells us that the world is made up of atoms. Cell theory tells us that living organisms are made up of cells. Boyle's law tells us that, as the volume of the lungs increases, their pressure decreases, which allows air to enter the lungs.

Anatomy and Physiology

Describe the types of information provided by anatomy and physiology

Anatomy (to cut up) studies biological structures and their relationships. It employs systematic observation, identification, and description. Typically, anatomy explores the structures of dead organisms (*dissection*). For example, anatomy gives names to the bones of the cranium: frontal, parietal, temporal, and occipital.

> The primary tools of the anatomist are observation and dissection
> Physiologists carry out experiments and develop theories

Anatomy answers the *what* questions, such as, What are the parts of the heart? *Physiology* (study of nature) concerns itself with the functions of the body. It explains how the anatomical parts work together to perform all the operations of the body. A physiologist carries out experimental investigations and provides theoretical explanations. Physiology typically studies living organisms (*vivisection*). It answers the *how* questions, such as, How does the heart pump blood? Experiments may be made in which a structure is removed or a drug is applied. To understand how the heart works, we might tie off the vagus nerve or inject an animal with epinephrine and observe that the heart beats faster in both cases.

Anatomists divide the body into organ systems and study how each contributes to the overall operation of the healthy body. Physiologists study the mechanisms by which life carries out its essential functions. Anatomy (observation) and physiology (experiment) are complementary sciences: structures carry out specific functions. For example, cranial bones are thick and closely spaced. They are ideal for protecting the brain.

Body Functions

List the major functions of the body

Physiology studies the basic functions of the body. *Body functions* may be divided into six categories. Living organisms:

1. are made of cells and products of cells organized to maintain a functional structure (*order*);
2. break down nutrients to provide energy and building blocks for new molecules (*metabolism*);
3. regulate their internal environment to maintain it in a relatively constant condition (*homeostasis*);

4. detect and respond to changes in their internal or external environment (*responsiveness*);

5. develop and increase in body size as they progress toward maturity (*growth and development*);

6. produce like kinds from generation to generation passing heritable characteristics to their offspring (*reproduction*).

> Living organisms maintain order, break down and utilize foods, maintain constant internal environment,respond to external environment, develop and grow to maturity, and produce offspring

In a biology course, we would add *evolution*, the adaptation of a population of organisms to changing conditions. However, traditional physiology is concerned with individual organisms and not populations. For this reason, evolution is not a subject of physiology.

Anatomical Systems

Describe how body structures are organized into functional systems

A living individual is an o*rganism*. This is the unit of study in physiology. Ideally, we want to understand how the body functions as a whole. When a patient comes into the clinic, we see the entire person and not a collection of cells or organs. Even so, we need to probe more deeply to find out what is wrong and usually begin a physical examination with a review of systems. These systems are the main anatomical divisions of the body.

Organ systems are collections of organs and related structures that work together. An *organ* is a distinct anatomical structure that carries out specific biological functions, such as the heart, lungs, or stomach. Vertebrates have eleven organ systems.

The *integumentary system* is the external body covering. It includes hair, skin, and nails. These structures isolate the body from the outside environment which is cold and dry, keeping the inside warm and wet. It also is the first line of defense against microbes and other pathogens.

Bones and joints make up the *skeletal system*. The skeletal system forms the framework of the body, giving it form and shape. It is the primary source of calcium, which is required for transmission of chemical signals, blood clotting, and contraction of the heart.

> Integumentary system is external body covering
> Bones and joints make up the skeletal system

Skeletal muscles and tendons form the *muscular system*. Muscles contract, pulling on tendons that are attached to the bones of the skeleton to cause movement. Cardiac muscles of the heart and smooth muscles that control blood vessel diameter, glandular secretions, and movement in the digestive tract are not part of the muscular system.

The brain, senses, and nerves make up the *nervous system*. The nervous system controls muscles and body organs. It is divided into the central nervous system (brain and spinal cord) and peripheral nervous system.

> Skeletal muscles and tendons form the muscular system
> Brain, spinal cord, senses, and nerves make up the nervous system

In the *endocrine system,* glands secrete hormones that act on body organs and other glands. The major endocrine glands are pituitary gland, thyroid gland, pancreas, adrenal gland, and gonads. Hormones are produced in one place in the body and act on another part of the body.

In the *cardiovascular system* the heart pumps blood through vessels that course throughout the body. Arteries pump blood away from the heart. Veins return blood to the heart. Arteries and veins are connected by capillaries that make a closed circulatory system.

> Endocrine system: glands secrete chemicals that act on organs
> Cardiovascular system: heart pumps blood through vessels

The *lymphatic system* collects extracellular fluid, removes pathogens, and returns lymph to the blood. Organs associated with the lymphatic system, thymus, lymph nodes, and spleen store and produce cells that mediate immune responses.

The *respiratory system* consists of the lungs and air passages that lead to them. Air enters and exits the lungs to supply oxygen to the blood and remove carbon dioxide.

> Lymphatic system collects extracellular fluid and returns it to blood
> Respiratory system consists of lungs and air passages that lead to them

The *digestive system* has organs that break down foods, absorb nutrients, and excrete solid wastes. Foods are broken down by chewing and mixing in stomach and small intestine and by digestive enzymes secreted by stomach, pancreas, and small intestines. The liver provides bile to aid digestion of fats.

In the *urinary system,* kidneys filter blood and its wastes are stored in the bladder for excretion from the body. About 20 % of the blood is filtered in the renal corpuscle, then filtered materials needed by the body, such as nutrients, water, and salts are reabsorbed by the tubules of the nephrons.

> Digestive system breaks down foods, absorbs nutrients, excretes wastes
> Urinary system: kidneys filter blood to form urine that is excreted

Gonads in the *reproductive systems* produce gametes that combine to continue the species. The male reproductive system produces sperm within the testes that pass through a tubule system to the join the urethra in the penis. Secretions in the tubular system add fluids to the sperm to produce semen.

In the female reproductive system, ova develop in ovaries and are released monthly to pass through the oviducts to the uterus where they are available to be fertilized by sperm implanted in the vagina. The uterine lining is replaced monthly.

> Reproductive systems produce gametes that combine to continue species

Tissues

List and distinguish among the four tissue types

Organs are made up of one or more tissues. *Tissues* consist of cells with similar structures and their extracellular matrix. Tissues demonstrate the principle that cells in an organism do not live in isolation. They are integrated into an extracellular matrix and surrounded by other cells within tissues. Tissues form the basic structural components of the body.

All tissues are classified into one of four types: epithelial, connective, muscle, and nervous. *Epithelial tissues* are mostly cells. They cover body surfaces, line body cavities, and form glands. *Connective tissues* are mostly extracellular matrix. They bind and support body parts, connecting one tissue with another. *Muscle tissues* are contractile, so move the body and its parts. *Nervous tissues* carry signals. They receive stimuli and conduct impulses. These four types of tissues are often found together in organs and major structures of the body.

> Epithelial cells cover body surfaces, line body cavities, and form glands
> Connective tissues (mostly extracellular matrix) bind and support body parts
> Muscle tissues are contractile; nervous tissues carry signals to specific sites

Biomolecules

Classify the molecules that carry out the functions of life

Anatomical structures are suited to their functions, so are molecules. Biological molecules are classified into four families which share common structural motifs. *Carbohydrates* (such as glucose) are important sources of energy and of recognition in the immune system. *Lipids* are used for energy storage, insulation, and protection, form the basic structure of cell membranes, and include steroid hormones. Fatty acids are lipids used for energy. Three fatty acids linked to glycerol form a *triglyceride*. *Nucleic acids* include energy transfer molecules (ATP), signaling molecules, and macromolecules (DNA and RNA) that direct protein synthesis. *Proteins* are the primary functional molecules. They serve as receptors, transporters, and enzymes. *Peptides* are small proteins.

Macromolecules are large biomolecules built up by linkage of simpler subunits. *Polysaccharides* are branched chains of simple sugars linked together. Glycogen is the storage form of the simple sugar glucose. DNA and RNA are linear chains of *nucleotides*. DNA forms a double helix. RNA is a single chain of nucleotides. Proteins are linear chains of *amino acids*. Proteins fold into complex functional structures.

> Polysaccharides are simple sugars linked together in branched chains
> Amino acids link together in linear chains to form proteins
> Nucleotides form single chains (RNA) or double chains (DNA)
> Triglycerides link three fatty acids to glycerol

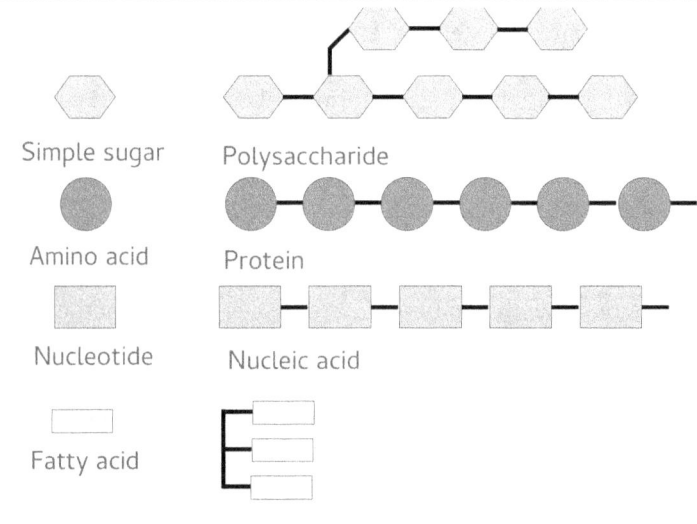

Simple sugar   Polysaccharide

Amino acid   Protein

Nucleotide   Nucleic acid

Fatty acid   Triglyceride

Homeostasis

Describe the concept of homeostasis and its role in preserving health

Living organisms must constantly adapt to changing conditions. *Homeostasis* preserves life and health in spite of large external variations. This is accomplished by maintaining a relatively constant *internal environment*. This internal environment is the fluid that bathes the cells that make up the living fabric of the body.

Conditions controlled by homeostasis include body temperature, blood pressure, and the pH and composition of blood. A *setpoint* is the target or desired value for a *controlled condition*, such as 37C for body temperature. Homeostatic control mechanisms have a *sensor* that monitors the controlled condition, a *control center* that compares the output of the receptor with a setpoint, and an *effector* that brings the controlled condition back toward the setpoint when it is disrupted.

A *control system* of homeostasis operates by feedback inhibition. The current state of the organism is constantly compared with a desired state. If a stimulus disrupts a controlled condition, control loops restore the disturbance to normal values. *Negative feedback systems* keep controlled conditions stable around a setpoint.

> Homeostasis maintains a relatively constant internal environment by keeping controlled conditions near an optimal setpoint
> Temperature is controlled within a narrow range by negative feedback

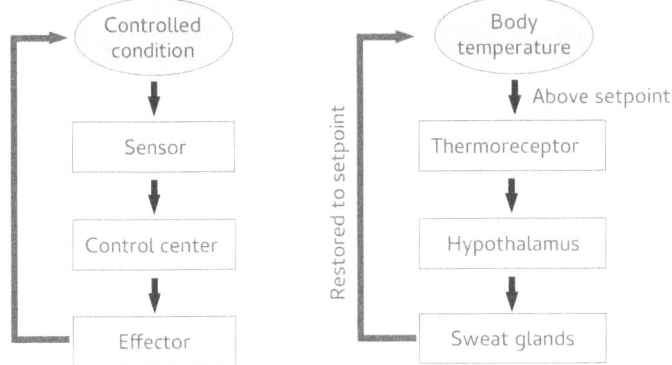

Thermoregulation

Explain the mechanisms that regulate and maintain body temperature

A good example of negative feedback is the control of body temperature. Body temperature is very closely regulated by negative feedback. The temperature control center is in the hypothalamus at the base of the brain, which monitors both blood and skin temperature. If the body is too warm, the hypothalamus responds by causing vasodilation of blood vessels in the skin and increased sweat production. Together, these increase heat loss and reduce body temperature. If body temperature drops, the response is heat conservation. Blood is diverted to the core of the body, we move and put on clothing to keep warm, and may start shivering.

> Body temperature is sensed by the hypothalamus which initiates response
> If too high, response is cutaneous vasodilation and perspiration
> If too low, response is sympathetic vasoconstriction and shivering

Thermoreceptors in the skin sense external temperature and provide a warning to the hypothalamus that the body has entered an adverse environment. This shell temperature is about 4C colder than the core temperature of body organs. Thus, skin thermoreceptors are responsible for the sensation of cold. When body temperature is below the setpoint in the hypothalamus, blood flow is diverted away from the body surface by vasoconstriction (narrowing of blood vessels). This not only reduces blood flow to the skin, it also shifts flow from surface veins to deeper veins. Heat loss through the skin is minimized and warm blood flow to core body tissues is maintained. Behavioral changes cause us to move to a warm place, slap our hands together to increase warming muscular activity, and put on more clothing for insulation. Shivering is a sympathetic response that increases heat output by skeletal muscles for short periods of time.

> Increased blood flow to skin causes loss of heat by radiation
> Vasoconstriction reduces blood flow to skin to preserve heart

In *fever*, the temperature setpoint is elevated, so we feel cold when body temperature is below the increased setpoint, even though temperature is above normal. *Pyrogens* (fever-producers) increase interleukin-1 (IL-1, an inflammatory mediator) that acts on the anterior hypothalamus to produce *prostaglandin E2* (PGE2), a fatty acid derivative that increases setpoint temperature. PGE2 is synthesized from arachidonic acid released from membrane phospholipids by the enzyme *COX-2* (cyclooxygenase type 2). Inhibition of COX-2 reduces fever by decreasing PGE2 production in the anterior hypothalamus. *Aspirin* inhibits COX (cyclooxygenase) enzymes, including COX-2. Like aspirin, *ibuprofen* is a non-specific COX inhibitor, so treats pain, fever, and inflammation. *Acetaminophen* is a non-aspirin pain reliever and antipyretic that specifically inhibits COX-2.

Review Questions

1. How does a scientist approach the study of the human body?
2. Give an example of how the scientific method might be applied to answer a question about body function.
3. Why is scientific knowledge always provisional?

4. How do the methods of anatomists and physiologists differ? How are their methods complementary?

5. List and define the six properties of life.

6. What is an anatomical system? Why are the systems of the body called "organ systems"?

7. List the eleven organ systems of the body and their primary functions.

8. What are the four basic tissue types? How are they distinguished?

9. What are the four classes of biomolecules? What does each do?

10. What is a macromolecule? Define polysaccharide, protein, and nucleic acid.

11. What is homeostasis and how is it important for health and well-being of the living organism?

12. What are the components of a homeostatic control system? Provide examples for each component.

13. Give an example of how negative feedback maintains a controlled condition near its setpoint.

Cell Biology

In the last lecture, we surveyed the eleven body systems and how they work together to maintain homeostasis and carry out the active functions of the body. We learned about tissues and biomolecules and control systems. In this lecture, we study the basic properties of cells. Living organisms are made of cells and the products of cells. The molecules produced by a particular cell is determined by which genes are active (*gene expression*). The genetic material is in the *nucleus*; *mRNA* carries that information into the cytoplasm, where *proteins* are made. The types of proteins made by a cell determine its functions. Cells are surrounded by a *plasma membrane* that isolates the extracellular fluids from intracellular fluids. Inside the cell, *organelles* carry out special functions, such as synthesis of proteins destined for secretion. Many of these organelles are surrounded by *cell membranes*.

Molecules are atoms linked together with covalent bonds. We will learn how to calculate and express their *concentrations* in solution. Energy for body processes is temporarily stored in the chemical bonds of *ATP*. Controlled breakdown (*metabolism*) of glucose and fatty acids by enzymes in the cytosol and mitochondria produce ATP. *Ions* are charged molecules or atoms that carry electrical charge (they are *electrolytes*). *pH* measures the concentration of the positively charged hydrogen ion. Close control of the pH surrounding cells is achieved by *buffer systems*, primarily by *bicarbonate buffer* in the extracellular fluids.

Cells

Describe the overall structure of the cell and its components

Cell theory tells us that the basic units of structure and function in all living organisms are *cells*. They are the smallest components of the body that carry out all the functions of life. Every cell has the same genetic information. Their different properties arise from differences in *gene expression*, the types of proteins produced by the cell. Since all cells arise from division of preexisting cells, these changes in gene expression arise by *differentiation* from an original fertilized egg (zygote).

Human cells have a cell membrane and nucleus easily visualized by light microscopy. The *cell membrane* entirely surrounds the cell, separating the *extracellular* region outside cells from its interior (*intracellular*). The *nucleus* contains the genetic material that determines the role of the cell in physiology. The region outside the nucleus is the *cytoplasm*.

> Cells are the basic units of structure and function in living organisms
> Cells (except red cells) have a distinct nucleus surrounded by cytoplasm
> Cell membranes separate intracellular and extracellular compartments

Organelles are distinct structures within cells that carry out particular functions. For example, ribosomes are organelles that synthesize proteins under the direction of mRNA. Some functional compartments in the cell are separated by membranes (*membranous organelles*).

The major membranous organelles are the nucleus, mitochondria, and endoplasmic reticulum. Mitochondria use oxygen to completely burn nutrients, producing large amounts of energy. The endoplasmic reticulum (ER) is a network of membranes that synthesize cholesterol and steroids (smooth ER) and proteins destined for secretion or the cell membrane (rough ER). The fluid part of the cytoplasm (that does not include organelles) is the *cytosol*.

Gene Expression

Explain how proteins produced by cells depend on what genes are expressed

The *central dogma* of molecular biology defines information flow in living organisms: *DNA* directs the synthesis of messenger RNA (*mRNA*) in the nucleus. mRNA enters the cytoplasm to direct the synthesis of proteins. The part of a DNA molecule that codes for protein is a *gene*. In *transcription* one strand (the template) of a gene guides the production of mRNA. In turn, triplets of bases (*codons*) direct the sequence of amino acids of new proteins during *translation*.

All cells are derived from one of three germ layers of the early embryo. These stem cells produce different cells by activating (*inducing*) and inhibiting (*repressing*) expression of particular genes. Similarly, some hormones and regulatory molecules alter gene expression, which changes the proteins made by the cell and their activity. Most of the DNA in the nucleus is not transcribed, but has structural and regulatory functions. The structure and function of a cell depends on the proteins it produces. This distribution is determined by which genes are active in producing mRNA that codes for those proteins (*gene expression*).

> Central dogma: DNA is transcribed to mRNA in nucleus
> mRNA enters cytoplasm to be translated into protein

Gene expression is regulated at each step of the process from DNA to protein. The first stage, which determines which gene will be transcribed, is controlled by *transcription factors*, which may induce or repress expression of the gene. Transcription factors bind to specific DNA sequences in multiprotein complexes. Small molecules that induce expression are *activators*. Those that decrease gene expression are *repressors*. Together, the complex of transcription factors, activator, and accessory (mediator) proteins recruits RNA polymerase to a specific site at the beginning of the gene. The result is more rapid transcription of the gene and increased synthesis of its protein product.

> Activator molecules bind to specific enhancer sites on DNA
> Binding brings mediator proteins into complex with transcription factors
> that bind to DNA promoter and activate transcription of gene

Cell Membrane Structure

Describe the structure of cell membranes

Cells are separated from their surroundings by a thin *cell membrane* that is *semipermeable*: some molecules may pass through the membrane into the cell; others are excluded. The basic structure of a membrane is a *phospholipid bilayer* with associated proteins. The same structure is found in internal membranes, so all are termed cell membranes.

Phospholipids consist of a glycerol backbone to which three components are linked: two of these are water-insoluble (*hydrophobic*) fatty acids and the third is a water-soluble (*hydrophilic*) head group. In a cell membrane, the hydrophilic phospholipid head group orients toward water and the hydrophobic fatty acid chains orient away from water. Cholesterol is buried within the fatty acid chain region of the phospholipids.

> Basic structure of cell membranes is phospholipid bilayer
> Polar groups orient toward water and fatty acid chains point away

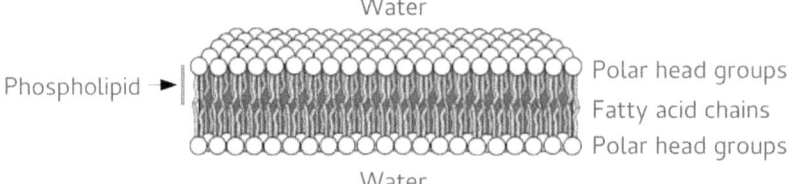

Phospholipids have a hydrophilic head group and two hydrophobic
fatty acid chains (tails) linked to a glycerol backbone

The hydrophobic fatty acid chains of phospholipids form the major barrier to passage of molecules across the membrane. Proteins embedded in the membrane are responsible for regulating passage of impermeable molecules across the cell membrane and responding to extracellular signaling molecules. These integral *membrane proteins* are classified as transporters, channels, and receptors.

Within the lipid bilayer of a cell membrane are proteins that serve
as channels, transporters, signaling molecules, and enzymes
Carbohydrates are attached to some membrane proteins and lipids

Transporters carry water-soluble materials across cell membranes. *Channels* carry ions across cell membranes. We will discuss their properties when we learn about excitable cells (neurons and muscle fibers). *Receptors* respond to binding of chemical messengers (hormones and neurotransmitters) by transmitting a signal into the cell. Molecules that bind to receptors are called *ligands*.

Molecules and Molecular Weight

Define molecule and describe how to calculate molecular weight

Matter is a material substance that takes up space. All matter is composed of atoms. *Atoms* have a dense central nucleus surrounded by a cloud of negatively charged electrons. The nucleus contains positive protons and neutral neutrons. Both have a mass slightly larger than 1 amu (atomic mass unit). The electrons surrounding the nucleus have almost no mass (about 1/2000 that of a proton or neutron).

Atomic number is the number of protons in the nucleus of an atom. Atoms of a single kind that share the same atomic number are elements. Elements are arranged according to atomic number in the periodic table. *Atomic mass* (sometimes called atomic weight) is approximately equal to the sum of protons plus neutrons.

Atoms are arranged in the periodic table by atomic number (of protons)
Atomic mass is nearly the same as the sum of protons and neutrons
Most common atoms in living organisms are in first 4 rows of periodic table

Atoms combine together by strong covalent bonds to form *molecules*. The *molecular weight* of a molecule is the sum of the atomic weights of its component atoms. For example, water has two atoms of hydrogen and one atom of oxygen. The atomic weights of hydrogen and oxygen can be found in a table of atomic weights or by referring to a periodic table. The atomic weight of hydrogen is 1.01 and that of oxygen is 16.00, so the molecular weight of water with two hydrogens and one oxygen (H_2O) is 18.02.

Solutions

Describe how to calculate concentrations of solutes in solution

A homogeneous mixture of substances is a *solution*. The dissolving agent in a solution is the *solvent*. The material that is dissolved is the *solute*. Water is a versatile solvent in which many biological molecules dissolve. The chemical bonds in water are *polar bonds*: its hydrogen atoms carry a slight positive charge and its oxygen is slightly negative. Water-soluble molecules are similarly polar. These slight charges allow a weak hydrogen bond to form between hydrogen and the slight negative charge of oxygen or nitrogen molecules on polar molecules. Hydrogen bonds to water make polar molecules soluble. Molecules that dissolve in water are *hydrophilic* (water-loving). Those that do not form hydrogen bonds and do not dissolve in water are *hydrophobic* (water-hating).

Hydrophilic (polar) molecules dissolve in water by forming hydrogen bonds between partially positive hydrogen and slightly negative oxygen or nitrogen atoms

The amount of solute dissolved in a solution (its *concentration*) is usually expressed as either percent or molar concentration. The amount of solute in grams dissolved in 100 mL of a solution is its *percent concentration* (%). Thus, 0.9 grams of NaCl in 100 mL of solution has a concentration of 0.9%.

Percent concentration is grams in 100 mL of solution Molar concentrations are gram-molecular weight in 1 L of solution

The molecular weight (molar mass) of a molecule is the sum of the atomic weights of its atoms. The molecular weight of NaCl is 23.0 (the atomic mass of Na) plus 35.5 (the atomic mass of Cl), which is 58.5.

Grams of a material divided by its molecular weight is its mass in *moles* (gram-molecular weight).

The mass of a solute in moles dissolved in one liter (L) of a solution gives *molar concentration* (with units of M). For a 0.9% solution of NaCl, first calculate the amount of salt in one liter (0.9 grams in 100 mL is 9 grams in 1 L). Then divide by the molar mass: 9/58.5 = 0.15. Thus a 0.9% NaCl solution is 0.15 M. Often, concentrations in the body are rather small, so are expressed in mM (1/1000 of M). A 0.15 M solution is 150 mM.

ATP

Describe the role of ATP in energy transfer in cells

Energy for muscle contraction is provided by *adenosine triphosphate* (ATP). ATP consists of adenine (a nitrogenous base), the sugar ribose, and three phosphates. *Adenosine diphosphate* (ADP) is identical to ATP but has only two phosphates. ATP is the energy currency of living systems. Addition of phosphate to ADP forms ATP which stores energy. Removal of phosphate from ATP forms ADP which releases energy. Energy released by conversion of ATP to ADP is harnessed to perform work, such as muscle contraction.

> When ATP is converted to ADP, energy is released to perform work
> ATP is produced as a result of controlled metabolic reactions

ATP is a temporary storage form of energy. Long-term storage resides in the chemical bonds of glycogen and triglycerides which can be split using water (by *hydrolysis*) to glucose and fatty acids that enter metabolic pathways in the cell to produce ATP. Fatty acids enter mitochondria to be broken down to produce energy. Glucose may either by metabolized to lactate in the cytoplasm or enter mitochondria to produce higher levels of ATP. No oxygen is required to produce lactate (anaerobic metabolism). Mitochondria require oxygen to carry out its metabolic reactions (aerobic metabolism).

Glycolysis and Anaerobic Metabolism

Describe the basic steps of glycolysis and production of lactate

Glucose (a simple sugar) is the major energy source for most cells and is required to produce energy in the central nervous system (brain and spinal cord). *Glycolysis* is the breakdown of glucose to two molecules of pyruvate in the cytoplasm. Glycolysis produces 2 ATP. Hydrogens from glucose are transferred to the nucleotide NAD to make NADH.

> Glycolysis converts glucose to pyruvate
> in the cytoplasm with net yield of 2 ATP

In the absence of oxygen, *anaerobic metabolism* converts pyruvate to *lactate* that leaves the cell to be recycled in the liver. NADH from glycolysis must be converted back to NAD to allow glycolysis to continue. This conversion is coupled to production of lactate from pyruvate by *lactate dehydrogenase*.

> Lactate dehydrogenase (LDH) produces lactate from pyruvate by
> converting NADH from glycolysis to NAD required in glycolytic pathway

Aerobic Metabolism

Explain how mitochondria produce ATP by oxidative phosphorylation

When oxygen is readily available, pyruvate enters mitochondria to be completely metabolized by aerobic metabolism to carbon dioxide and water with production of large amounts of ATP. Because it requires oxygen, this is called *aerobic metabolism*. Fatty acids also enter mitochondria to produce ATP by aerobic metabolism.

> Acetyl-CoA for citric acid cycle in mitochondrial matrix arises from pyruvate
> with release of carbon dioxide or beta-oxidation of fatty acids

The central pathway in the mitochondrial matrix is the citric acid cycle which takes place in the fluids of the mitochondrial matrix inside the inner membrane of the organelle. Pyruvate and fatty acids are first converted to acetyl-CoA, then enter this cycle which produces ATP and the reduced nucleotides NADH and FADH2. The breakdown of fatty acids into two-carbon acetyl-CoA units is *beta-oxidation*.

Mitochondria produce energy in cells by oxidative phosphorylation of ADP to ATP. Mitochondria have a double (inner and outer) membrane. The inner mitochondrial membrane is folded into cristae studded with enzymes. Enzymes in the mitochondrial inner membrane carry out oxidative phosphorylation (converting ADP to ATP). Aerobic metabolism yields about 30 ATP per glucose molecule. Aerobic metabolism is more efficient, but much slower than glycolysis.

> Citric acid cycle produces carrier molecules that enter
> electron transport chain that produces H ion gradient
> ATP is produced from ATP as H ions pass through ATP synthase

NADH and FADH2 electrons drive protons across the inner mitochondrial membrane (electron transport). When the protons (H ions) pass back across the membrane, they drive phosphorylation of ADP to ATP.

Each NADH yields 3 ATP and each FADH2 yields 2 ATP during oxidative phosphorylation. The electrons are transferred to oxygen along with two protons to produce water.

Cations and Anions

Distinguish between anions and cations

When a salt, such as KCl is dissolved in water, it dissociates into ions. *Ions* are atoms or molecules that carry a net charge. *Cations* are positively charged ions. They are represented by writing a + sign, such as K^+. *Anions* are negatively charged ions, for example, Cl^-.

> Salts dissociate in water to form charged ions
> Positively charged ions (Na+, K+) are cations
> Negatively charged ions (Cl–) are anions

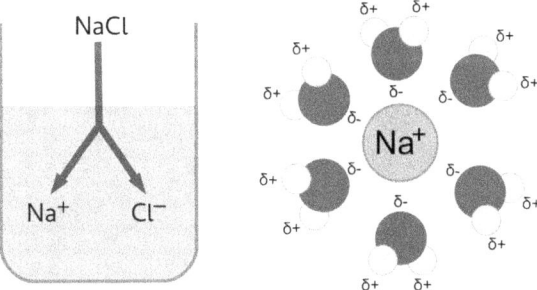

Ions conduct electricity in water solutions because of their charge. For this reason, they are called *electrolytes*. Molecules, such as glucose, do not dissociate into ions in water so do not conduct electricity. However glucose and ions hydrogen bond to water, so both are surrounded by solvation shells of water that make them large in solution.

pH

Distinguish between acids and bases and describe the pH scale

pH is a measure of the concentration of hydrogen ions in solution. Tight control of pH is critical to life and health. Even small changes in pH can have dramatic effects on metabolism. Formally, pH is defined as the negative log (the exponent) of the molar hydrogen ion concentration, $pH = -\log [H^+]$, when concentration is expressed as moles (molecular weight in grams) per liter. Thus, a neutral solution has a pH of 7. *Acids* have more H^+ (a greater $[H^+]$) and a pH < 7. *Bases* have less H^+ (a smaller $[H^+]$) and a pH > 7. pH is an inverse log scale: a solution of pH 3 has 10X as many hydrogen ions as one of pH 4 and 100X as many hydrogen ions as one of pH 5.

> Water dissociates into small, but equal numbers of
> hydrogen ions (acid and hydroxyl ions (base)
> Concentration of hydrogen ions gives pH = $-\log[H^+]$

A small part of water dissociates into an equal number of hydrogen cations and hydroxyl anions. Pure water has a pH of 7. Acids (such as HCl) release hydrogen ions into solution, increasing the hydrogen ion concentration. This explains their low pH. Bases (such as NaOH) take up hydrogen ions from solution, decreasing $[H^+]$ which means they have an elevated pH.

Buffers

Explain the role of buffers in maintaining constant body pH

> Buffers maintain a relatively constant pH with addition of acids or bases
> Addition of acid (such as HCl) increases hydrogen ion concentration
> causing decreased pH in water and minimal change of pH in buffer

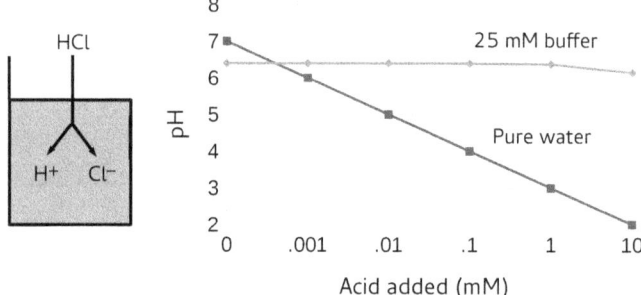

Acid added to water alone rapidly decreases pH as more acid is added. Similarly, addition of a base to water increases its pH. Neither is ideal in the body. The body must maintain a nearly constant pH, even when metabolic acids are being produced. This is accomplished by *buffer systems*, which maintain a relatively constant pH even when acids or bases are added.

Bicarbonate Buffer System

Explain how the bicarbonate buffer system maintains constant pH in extracellular fluids

A *buffer* is a mixture of a weak acid (or weak base) and its salt. For the *bicarbonate buffer system*, carbonic acid is the weak acid and the salt is bicarbonate. A weak acid does not fully dissociate in solution, unlike a strong acid (HCl, for instance) which fully dissociates.

Bicarbonate buffer system is primary buffer in plasma
Addition of acids to bicarbonate produces carbonic acid
Removal of hydrogen ions by base addition produces bicarbonate

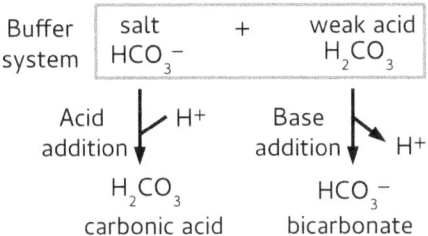

Bicarbonate is produced by loss of H^+ from carbonic acid, so it carries a negative charge. This negative charge is balanced by a positive charge from sodium or potassium ions in the plasma to form a salt. Buffers are most effective at their pKa, the pH at which the concentration of weak acid (or base) and its salt is equal.

Review Questions

1. What is cell theory? What insight does the study of cells provide for the student of physiology?

2. Distinguish extracellular from intracellular and cytoplasm from cytosol. What is a membranous organelle?

3. What is meant by gene expression? What is the central dogma?

4. What is the difference between transcription and translation? In what compartment of the cell does each take place?

5. What is the role of mRNA in protein synthesis?

6. What is the basic structure of a cell membrane? Describe the structural features of a phospholipid.

7. What are the functions of protein molecules in cell membranes?

8. The molecular formula for glucose is $C_6H_{12}O_6$. Using an atomic weight of 12 for carbon, 16 for oxygen, and 1 for hydrogen, calculate the molecular weight of glucose.

9. Describe how to prepare a 5% glucose solution. What would be the molarity of this solution?

10. When a salt is dissolved in water it dissociates into what two components? Why are salts called electrolytes?

11. Describe the role of ATP in transferring energy.

12. How much ATP is produced by glycolysis? Where do its reactions take place?

13. What is the end product of glycolysis? How is lactate formed?

14. What are the sources of acetyl-CoA for the citric acid cycle?
15. How are fatty acids broken down and where does this occur?
16. How is the citric acid cycle linked to production of ATP?
17. What is oxidative phosphorylation and where does it take place?
18. What is pH? Why is the pH of pure water equal to 7? Distinguish between acids and bases.
19. What is a buffer and what is the effect of adding acids and bases to a buffer system?
20. How does the bicarbonate buffer system work?

Visceral Control

The most beautiful and elevated problem for the human intellect, the discovery of the laws of vitality, cannot be resolved, nay, cannot even be imagined, without an accurate knowledge of chemical forces.

– Justus von Liebig

Messengers and Receptors

You have gained some understanding of how cells carry out their functions and the basic properties of molecules and ions. Now, we turn to the molecular mechanisms that explain how the body carries out its functions. Students often find molecular mechanisms challenging because it is easier to visualize an anatomical structure and its function than a molecular one. However, the topic is an important one. Medications work by binding to *receptors*.

Natural molecules that bind to receptors are *chemical messengers*. Receptors respond to their binding by changing the activity of cells and, thereby, the functional state of the body. *Hormones* are messengers that travel through the blood to act on tissues distant from their site of production. *Neurotransmitters* are messengers released by nerve endings that are directed to specific sites in the body. Some messengers bind to tightly to receptors (have high *affinity*) so cause effects at low concentrations. Others have lower affinity. Chemical messengers can be classified by their solubility. Water-soluble messengers do not enter cells but act on *cell surface receptors* that carry signals across the plasma membrane. Hydrophobic molecules, such as steroids, enter cells to bind to *nuclear hormone receptors* that modify gene expression.

Chemical Messengers

Explain how hormones and the autonomic nervous system regulate body organs (viscera)

The activity of body organs and tissues (viscera) is controlled by two body systems that act together to maintain homeostasis: the autonomic nervous system and the endocrine system. The *autonomic nervous system* (ANS) sends neurons to specific sites, where *neurotransmitters* (chemical messengers) are released that alter the activity of the body organ or tissue. An example is the vagus nerve which reduces heart rate when activated. The *endocrine system* produces hormones that enter the blood to pass throughout the body. *Hormones* are chemical messengers produced by endocrine cells that cause an effect on other organs and tissues.

> Neurotransmitters and hormones bind to receptors to change cell function
> Neurons travel directly to target tissue where neurotransmitter is released
> Hormones enter circulation to bind to specific receptors on cells

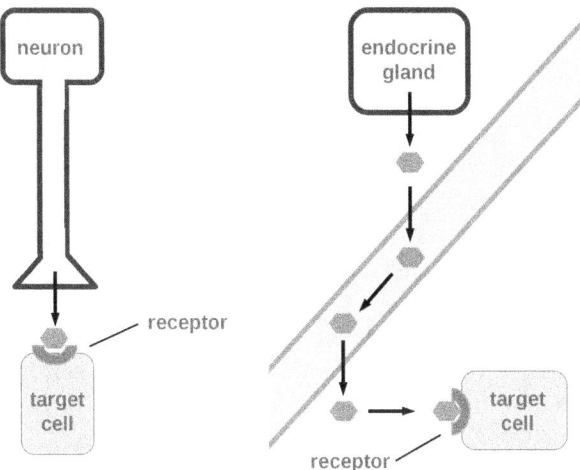

The endocrine and autonomic nervous systems function together to regulate body activities. Most organs are influenced by both systems. Both cause cellular responses because chemical messengers bind to specific *cellular receptors*. In the ANS, the chemical messengers are neurotransmitters released at *synapses* from neurons that go to specific target organs. Thus, they can act quickly and locally. Hormones are produced in *endocrine organs*. They must travel through the blood to reach their target organs and tissues. There they act only on *target cells* that have receptors for that specific hormone. Generally, hormones influence longer-term body functions, such as growth, metabolism, and reproduction.

Cell Receptors

Describe the role of receptors for chemical messengers

Chemical messengers include hormones and neurotransmitters. Hormones are secreted by endocrine glands, enter the blood, and travel to other tissues, where they exert their effects. Closely related to hormones are *paracrines*, signaling molecules that act on nearby tissues, and *autocrines*, which act on the cell that produces them.

Neurotransmitters are released from axon terminals of neuronal synapses. They pass across a narrow gap (synaptic cleft) between the axon terminal and a post-synaptic cell where they act upon cell membrane receptors. These receptors are usually on the postsynaptic cell which results in transmission of the signal from one cell to another. The postsynaptic cell may be another neuron, a gland, or a muscle cells. Some neurotransmitters act on the presynaptic cell to inhibit secretion of neurotransmitter.

> Autocrine signals act on the cell that secretes them
> Paracrine signals act on neighboring cells
> Endocrine signals enter blood to act on distant tissues

Chemical messengers act on cells by binding to *receptors*. Receptor binding changes cellular activity. The term receptor has a broad meaning in pharmacology. It refers to any molecule that causes functional changes in the body when another molecule binds to it. Cellular receptors receive a signal from the chemical messenger and carry it into the cell, which changes its function in response. The response depends on the chemical messenger and the specific receptor to which it binds.

> Chemical messengers (hormones and neurotransmitters) bind
> to cellular receptors to cause a response in the cell

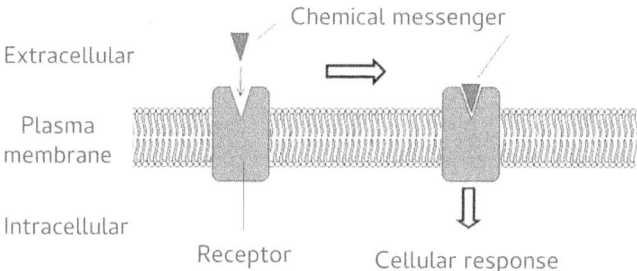

Each type of cell has a distinct set of receptors that determine how it will respond to extracellular signals. Receptors are either buried in the cell membrane with the receptor site exposed to the extracellular fluids (*cell surface receptors*) or inside the cell (*intracellular receptors*). Messengers do not need to enter the cell to act on cell surface receptors. They must enter the cell to bind to intracellular receptors.

Receptor Binding

Distinguish between high and low affinity ligand binding to a receptor

Biological functions are carried out by protein receptors, transporters, and enzymes. *Receptors* bind specific molecules (*ligands*) to alter cellular function. *Enzymes* bind specific molecules (*substrates*) at their active sites to carry out a reaction that produces a *product*. Transporters and channels carry specific molecules across cell membranes. Any of these classes of functional molecule may be the site of action of a medicinal drug, so pharmacologists refer to all of them as receptors.

As a result of binding of a substrate to the active site of an enzyme, the reaction proceeds faster than it would in its absence. For example, addition of water to sucrose (hydrolysis) is carried out on the active site of the enzyme sucrase (enzymes are often named for their substrates, with the suffix -ase if they cause hydrolysis). Binding strains the linkage between the two simple sugars which results in cleavage of the bond between them and release of the products glucose and fructose.

Substrates bind to active sites of enzymes where reaction takes place
Products are released after reaction

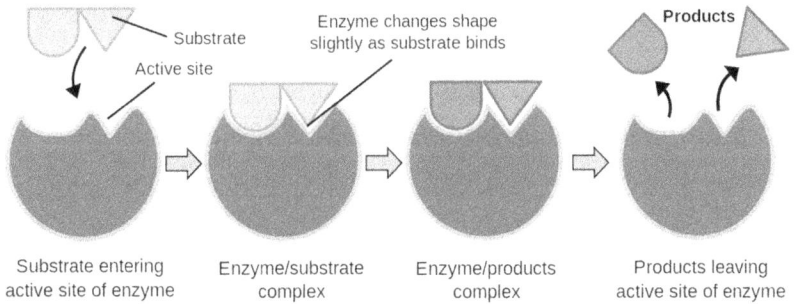

| Substrate entering active site of enzyme | Enzyme/substrate complex | Enzyme/products complex | Products leaving active site of enzyme |

Hormones, neurotransmitters, and most drugs act on specific receptors to alter biological functions. Like substrate binding to the active site of an enzyme, ligand binding to a receptor site is highly specific. The structure of the ligand is complementary to that of the receptor, which changes shape slightly as the ligand binds. *Agonists* duplicate the activity of a natural ligand. *Antagonists* compete with the natural ligand for binding to the receptor. A molecule that stops an enzymatic reaction is an *inhibitor*.

Agonist binding causes cellular response similar to natural ligand
Antagonists block receptor binding of natural ligand

Affinity is the strength of binding of a ligand to its receptor. A ligand is said to have *high affinity* for a receptor if low concentrations are effective in causing a response. *Low affinity* ligands must be present in much higher concentrations to cause a response. Binding affinity is compared using *EC50s* (effective concentration at 50%), the concentration of ligand required for half the maximal receptor effect. High affinity ligands have low EC50s; low affinity ligands have high EC50s.

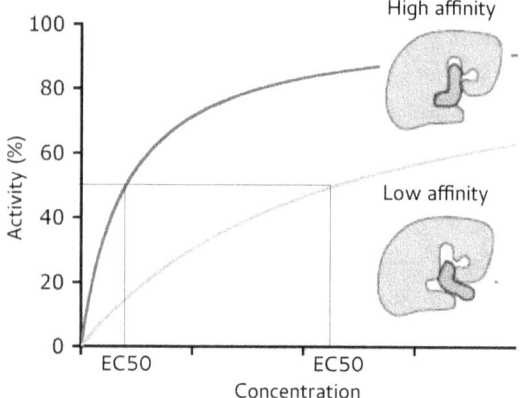

High affinity ligands bind to receptors at low concentrations
Low affinity receptors bind only at relatively high concentrations

Hydrophilic and Hydrophobic Ligands

Distinguish between water-soluble and hydrophobic ligands

Proteins, peptides, and catecholamines (such as epinephrine) are water-soluble. Water-soluble ligands travel free in plasma and do not need to bind to any carrier. Because they are water-soluble, they do not pass through cell membranes, but bind to protein receptors on the cell surface (*cell surface receptors*).

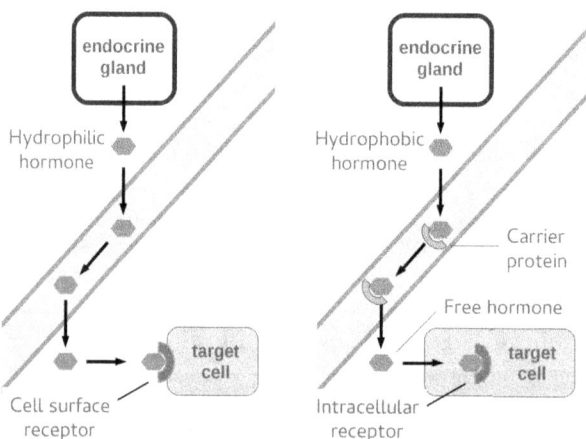

Hydrophilic hormones dissolve in plasma and bind to cell surface receptors
Hydrophobic hormones require carriers and bind to intracellular receptors

Steroids and thyroid hormones are insoluble in water (are hydrophobic). They are transported in plasma by carrier proteins. The carrier proteins remain in blood vessels. The hormone is released from the carrier protein, enters extracellular fluids, and directly passes through cell membranes to bind to *intracellular receptors*.

Cell Surface Receptors

Describe how water-soluble hormones alter cellular activity

Binding of a water-soluble hormone to its *cell surface receptor* initiates a cascade of reactions within the cell. In most cases, this *signal transduction* involves second messenger molecules that alter cellular function. (The first messenger is the ligand binding to the cell membrane receptor, but it is never called that.)

> Water-soluble hormones bind to cell surface receptors
> initiating a signal transduction cascade that alters cellular function

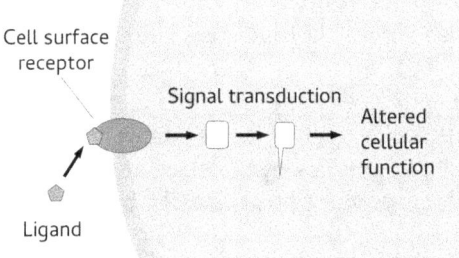

Signal transduction pathways can be rather complex and vary greatly from one ligand-receptor complex to another. Often, signal transduction begins with activation of a small, cell surface G-protein that is coupled to the receptor. Such receptors are GPCRs (*G-protein coupled receptors*) and are the most common biological target of drugs. Small G_s proteins (S for stimulatory) activate the intracellular enzyme adenylyl cyclase, which converts ATP to cAMP (cyclic AMP). cAMP acts as the second messenger by activating protein kinases that phosphorylate critical regulatory proteins in the cell. Phosphorylation activates some proteins and inhibits others. This shifts the cell into a different functional state.

> Many water-soluble ligands bind to GPCRs coupled to G proteins inside cell
> G proteins activate signal transduction cascades (often through cAMP)
> Second messengers alter activity of regulatory enzymes

Nuclear Hormone Receptors

Describe the signaling mechanisms of lipid-soluble hormones

Hydrophobic hormones are divided into two structural classes: lipid derivatives and amino acid derivatives. Lipid derivatives are derived from cholesterol are steroids; those derived from fatty acids are eicosanoids, of which prostaglandins are a subset. Thyroid hormones are amino acid derivatives. Since they are iodinated, they are not water-soluble.

All hydrophobic hormones must be carried in plasma bound to transporters. Hormones must first be released from the transporter before they can pass through the cell membrane and enter the cell. Within the cell *hydrophobic hormones* bind to *intracellular receptors* that alter gene expression. Since intracellular receptors act on genes in the nucleus, they are often called *nuclear hormone receptors*.

> Lipid-soluble (hydrophobic) hormones directly enter cells
> to bind to intracellular receptors that enter the nucleus
> Nuclear hormone receptors alter gene expression and cellular activity

In the cell, hydrophobic hormones bind to *intracellular receptors*, either in the cytosol or in the nucleus. The complex of hormone and receptor acts within the nucleus where it influences DNA transcription. Some genes are activated while are others are inhibited. This changes the pattern of mRNA produced in the nucleus (gene expression). The end result is that the proteins synthesized in the cytoplasm is altered. Altering the proteins produced by the cell changes cellular function.

Review Questions

1. What are autocrine, paracrine, and endocrine messengers?
2. What is a receptor? What kinds of molecules bind to receptors?
3. How are enzyme substrates similar to ligands that bind to receptors?
4. How is an agonist different from an antagonist?
5. What is the difference between a low affinity and high affinity ligand? Define EC50.
6. How are hydrophilic and hydrophobic hormones carried in blood? To what kinds of receptors do they bind?
7. How do hormones use second messenger systems to alter target tissue function? What is a GPCR?
8. How do steroid and thyroid hormones regulate cell function at target tissues? What is a nuclear hormone receptor?

Endocrine System

When hormones bind to receptors, they alter cellular function. Binding of hormones to cell membrane receptors causes a change in the receptor structure that initiates a series of events (signal transduction) that leads to a change in cellular enzyme activity. Intracellular nuclear hormone receptors alter gene expression. In the next two lectures, we will look into the two control systems of the body that regulate organ function: the *endocrine system* and the autonomic nervous system. Both systems are regulated by lower brain centers in the *hypothalamus* and brain stem. Many endocrine systems are controlled by hormones of the *anterior pituitary gland* that act on other endocrine organs (*tropic hormones*). In the current lecture, we will look at control of *thyroid gland* secretions and the hormones of secreted by the *adrenal cortex* as examples of hormonal control.

Endocrine Organs

Describe the components and functions of the endocrine system

The *endocrine system* has glands that secrete chemicals (hormones) that act on other organs. *Hormones* are secreted by *endocrine glands* into the surrounding extracellular fluid where they pass into the blood to be carried to all parts of the body. Endocrine glands thus differ from exocrine glands, such as sweat and oil glands, that secrete chemicals through ducts onto epithelial surfaces. The major endocrine glands are pituitary, thyroid, and adrenal glands, pancreas, and gonads.

> Hormones are produced by glands located at specific sites in the body
> Hormones act at sites distant from the glands that secrete them
> Hormones regulate growth, metabolism, and reproduction

Endocrine control is due to chemicals (hormones) that travel all over the body. The responses to hormones are of slower onset and longer-lasting than those in the nervous system. Thus hormones act on organs to control long-term physiological processes, such as reproduction, growth, and metabolism. Even though hormones circulate throughout the body, they are highly specific. Only cells with receptors for particular hormones respond to their influence.

Hypothalamus

Describe the role of the hypothalamus in the endocrine system

The *hypothalamus* is the major integrating link between the nervous and endocrine systems. This brain structure is located anterior to the brainstem at the base of the frontal lobe. The hypothalamus directly controls body temperature, blood pressure and volume, and pH by regulation of the activity of the autonomic nervous system (ANS).

> Hypothalamus is anterior and inferior to thalamus and connects to pituitary gland through infundibulum (stalk)

The hypothalamus is intimately linked with the pituitary gland, which is part of the endocrine system. The *pituitary gland* is located inferior to the hypothalamus to which it is connected by the infundibulum (stalk). It is protected by the sella turcica of the sphenoid bone, which fully surrounds the gland. The anterior lobe arises from an invagination of the oral ectoderm, while the posterior lobe is of neural origin.

In addition to its role in regulation of basic body functions through the ANS, neurons in the hypothalamus produce hormones that are directly secreted into the posterior pituitary gland or into nearby capillaries. The latter hypothalamic hormones pass through the *hypophyseal portal blood vessels* of the infundibulum to the anterior pituitary.

Neurosecretory Cells

Describe how hormones are secreted from neurons into the blood

The hypothalamus is part of the nervous system. It produces hormones in *neurosecretory cells* that secrete the hormones into interstitial fluids, where they enter blood vessels. In a neurosecretory cell, chemical messengers (hormones) are synthesized in the cell body where the nucleus is located, travel down the single axon to axon terminals, and are secreted into extracellular fluids at synapses. There the secreted hormones enter nearby capillaries to be distributed to target cells through the blood.

Neurosecretory cells have cell bodies in the hypothalamus. Hormones destined for the anterior pituitary gland are secreted locally, in the hypothalamus, where the enter the portal circulation of the infundibulum. Those destined for the posterior pituitary send their axons through the infundibulum and terminate in the gland, where hormones are released.

> Hormones produced in cell bodies of neurosecretory cells are transported along axons to axon terminals and released from synapses into extracellular fluids, then diffuse into blood

Posterior Pituitary Hormones

List the hormones of the posterior pituitary gland and their functions

The peptide hormones oxytocin and ADH are synthesized by neuroendocrine cells that arise in the hypothalamus and travel to the *posterior pituitary gland* where the hormones are secreted into blood. Each hormone arises from a different region of the hypothalamus. Oxytocin is synthesized in the supraoptic nucleus (directly above the site where optic nerves pass beneath the brain). ADH is synthesized in the paraventricular nucleus (adjacent to the third ventricle that surrounds the thalamus in the center of the brain, just beneath the cerebrum).

Both hormones have direct effects on the body. *Oxytocin* induces uterine contractions in response to cervical stretching during labor and causes contraction of myoepithelial cells of the mammary glands (milk ejection) when the nipples are stimulated. It also plays a role in social bonding with sexual partners and maternal behaviors. ADH (*antidiuretic hormone*) is also called vasopressin due to its effects on blood vessels when injected at high concentrations. ADH increases water retention by the kidneys when plasma osmolarity increases (typically, due to blood loss or dehydration). Because ADH decreases urine volume, it is an antidiuretic. This effect preserves body water, which increases blood volume to help maintain normal blood pressure.

> Oxytocin is synthesized in supraoptic nucleus of hypothalamus
> ADH (anti-diuretic hormone) is synthesized in paraventricular nucleus
> Both OT and ADH are released into circulation in posterior pituitary

Tropic Hormones

Explain the concept of tropic hormones and give an example

Hypothalamic hormones that are secreted into the blood and travel through the hypophyseal portal vessels to the anterior pituitary gland regulate secretion of hormones by the anterior pituitary. None have direct effects on the body: they are regulatory hormones. *Releasing hormones* increase release of a particular hormone from the anterior pituitary and release-inhibiting hormones decrease release of specific hormones. Nearly all hypothalamic hormones are releasing hormones.

Most of the hormones secreted by the anterior pituitary gland similarly do not have direct effects on the body, but exert their effects by stimulating release of hormones from other endocrine glands. Such hormones are generally called *tropic hormones*. The terminology is distinct from that of *trophic* hormones, which have growth-promoting effects and are often tumor promoters. Tropic means attraction or a turning. Trophic means nourishment. Unfortunately, the two terms are often incorrectly used interchangeably.

> Hypothalamus produces releasing hormones that enter anterior pituitary to cause release of tropic hormones that act on other endocrine glands which secrete hormones that act on body organs and glands

Anterior Pituitary Hormones

List the hormones released by the anterior pituitary gland

The hypophyseal portal system carries releasing hormones in blood from the hypothalamus to the *anterior pituitary gland* to control secretion of hormones that act on other endocrine glands. Hypothalamic hormones may either stimulate or inhibit release of anterior pituitary hormones. Most hypothalamic hormones are releasing factors (hormones) that stimulate secretion of tropic hormones from the anterior pituitary. A tropic hormone stimulates release of hormones from another endocrine gland.

> Neurosecretory cells of hypothalamus secrete hormones into hypophyseal portal circulation to enter anterior pituitary Hormones stimulate or inhibit secretion of anterior pituitary hormones

The names of the tropic hormones are rather long, so are typically abbreviated. In the following table are summarized the relationships between hypothalamic releasing hormones, anterior pituitary tropic hormones, and systemic hormones that act on the body.

Hypothalamus	Ant. Pituitary	Target Organ	Hormone
TRH	TSH	thyroid gland	T4 and T3
CRH	ACTH	adrenal cortex	cortisol
GnRH	FSH and LH	gonads	sex hormones
GHRH	GH	liver	IGF1

TRH: thyrotropin-releasing hormone
CRH: corticotropin-releasing hormone
GnRH: gonadotropin-releasing hormone
GHRH: growth-hormone releasing hormone
TSH: thyroid-stimulating hormone
ACTH: adrenocorticotropin hormone
FSH: follicle-stimulating hormone
LH: luteinizing hormone
GH: growth hormone

Some anterior pituitary hormones act directly on organs in the body. FSH (follicle-stimulating hormone) and LH (luteinizing hormone) have direct effects on development of ova and sperm, in addition to stimulating secretion of sex hormones. GH (growth hormone) stimulates systemic growth and stimulates secretion of growth-stimulating IGF-1 (insulin-like growth factor) by the liver. Prolactin acts on mammary glands to stimulate milk production (while oxytocin from the posterior pituitary stimulates milk ejection).

Thyroid Hormones

Describe the control of thyroid hormone synthesis and release

The *thyroid gland* is located in the neck, inferior and lateral to the larynx. Most of the gland is taken up by thyroid follicles. A thyroid follicle consists of a set of cuboidal epithelial cells (*thyroid follicular cells*) surrounding an extracellular depot (colloid). These cells synthesize the protein thyroglobulin and package it into membrane-bound vesicles in the rough endoplasmic reticulum.

Follicular cells of thyroid gland produce thyroglobulin protein
Dietary iodide enters blood and is transported into colloid by follicular cells
Iodide links to thyroglobulin producing storage form of thyroid hormones

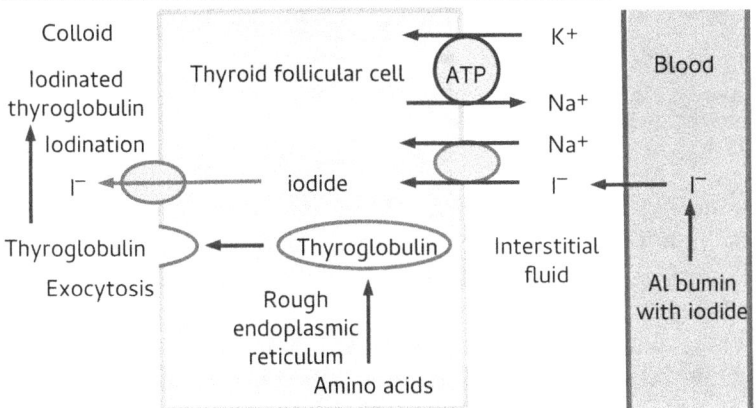

Thyroglobulin is secreted by *exocytosis* into the colloid. Dietary iodide enters follicular cells and passes into the colloid where it is oxidized to iodinium. This form of iodine reacts with tyrosine residues of thyroglobulin to produce iodinated tyrosine residues. Two nearby modified tyrosines link together to form the precursor of thyroid hormones, which remain part of the protein. In this way, the hydrophobic thyroid hormones are stored as iodinated *thyroglobulin* that can be mobilized as needed.

Stimulation of follicular cells by TSH from the anterior pituitary gland causes uptake of thyroglobulin into the follicular cell by endocytosis.

Endocytosis brings materials into a cell and causes them to fuse with small membranous organelles (*lysosomes*) that contain digestive enzymes. Lysosomal digestion of thyroglobulin releases the hormones. Nearly all of the hormone released is T4 (*thyroxine*), which has 4 iodines. Only a small amount of T3 (*triiodothyronine*), with 3 iodines, is released. Most of the highly active T3 is produced in peripheral tissues by deiodination of T4.

> TSH stimulates endocytosis of thyroglobulin which is digested by lysosomes to release thyroid hormones (T4) that are secreted into interstitial fluids
> Thyroid hormones enter blood and bind to thyroxine binding globulin (TBG)

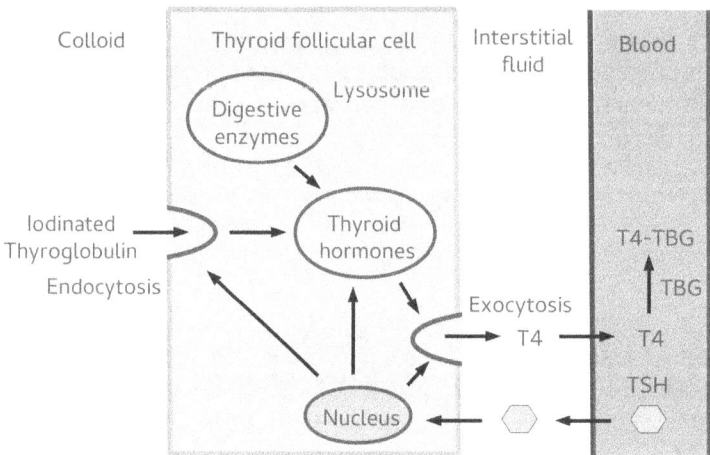

Thyroxine enters the blood and is transported throughout the body bound primarily to the protein *thyroxine-binding globulin* (TBG). Thyroxine must be released from its binding protein to enter target cells where it binds to intracellular thyroid hormone receptors that modify gene expression. Thyroid hormones increase basal metabolic rate by stimulating breakdown of glucose and fatty acids for ATP production, growth, and development. Low thyroid hormone levels (hypothyroidism) are associated with tiredness and developmental delays.

Regulation of Thyroid Hormone Levels in Blood

Explain how thyroid hormone levels are regulated and common thyroid hormone imbalances

Blood concentrations of thyroid hormones and TSH remain relatively constant throughout the day. Feedback is by the hormones themselves. Hypothalamic TRH stimulates release of TSH from the anterior pituitary gland. TSH stimulates secretion of thyroid hormones from thyroid follicular cells. At the level of the tissues T4 is converted to the more active T3 by an iodinase. Both T3 and T4 feed back to inhibit release of TRH and TSH.

The most common cause of *hypothyroidism* is iodine deficiency, which may be treated by providing iodine in the diet. When dietary iodine is sufficient, the most common cause of hypothyroidism is *Hashimoto's thyroiditis*, an autoimmune disease of the thyroid gland that makes it non-functional. It is treated with *levothyroxine*, a synthetic T4, that is the most commonly prescribed medication in the United States. Levothyroxine has a narrow therapeutic index. Regular thyroid function tests are required to make ensure the dose is adequate but not so high that it causes adverse effects, such as anxiety and heart palpitations. The most common lab tests are for TSH and free T4 (not bound to thyroxine-binding globulin).

> TRH secreted by hypothalamus stimulates secretion of TSH from anterior pituitary gland which acts as a tropic hormone on thyroid gland to stimulate release of thyroxine that is converted to T3 in periphery

If iodine is not available to synthesize thyroid hormone precursors, no hormone is produced. These low thyroid hormone levels increase TRH and TSH release. Increased TSH causes the gland to increase in size in an attempt to overcome its failure to produce hormones. The resulting enlarged thyroid gland is often visible as a swelling in the neck called *goiter*.

Grave's disease is the result of autoantibodies that mimic TSH in stimulating growth of the thyroid gland. The result is an enlarged thyroid that overproduces thyroxine. A common sign is a characteristic and irreversible bulging of the eyes. High levels of thyroxine suppress production of TRH and TSH, whose function is replaced by the autoantibodies. Grave's disease is treated with radiation therapy to destroy the thyroid gland. Hormonal replacement therapy with levothyroxine is required to restore normal levels of thyroxine.

Growth Hormones

Describe the hormones that stimulate systemic growth

Hypothalamic GHRH (*growth hormone releasing hormone*) stimulates release of GH (*growth hormone*) from the anterior pituitary gland. Growth hormone has anabolic effects: it stimulates growth and increases muscle mass. In addition, GH acts as a tropic hormone: the liver secretes *insulin-like growth factors* (IGF-1) in response to increased levels of growth hormone. Hepatic IGF-1 increases amino acid uptake and protein synthesis. It has a molecular structure similar to insulin, thus its unusual name. Peak levels are associated with a growth spurt during puberty.

> Hypothalamic GHRH stimulates GH release from anterior pituitary gland
> GH (growth hormone) stimulates growth and is a tropic hormone
> that stimulates secretion of insulin-like growth factor (IGF-1) by the liver

Laron dwarfism is due to a deficiency of hepatic GH receptors which activate IGF-1 synthesis. The syndrome does not respond to growth hormone replacement therapy. Excess growth hormone before puberty causes accelerated growth, leading to *gigantism*. In adults (whose epiphyseal plates have closed), excess growth hormone results in *acromegaly*, a disfiguring enlargement of hands, feet, forehead, and jaws. Typically, it is the result of a pituitary tumor.

Adrenal Cortex

Describe the hormones secreted by the adrenal cortex and their regulation

The *adrenal glands* sit on top of each kidney. Each gland has an outer cortex that secretes steroid hormones and an inner medulla that secretes epinephrine. The *adrenal cortex* (outer layer) primarily produces two types of steroid hormones: mineralocorticoids and glucocorticoids. *Mineralocorticoids* (primarily aldosterone) are produced in response to angiotensin II to conserve water in the kidneys, producing a more concentrated urine and increasing blood volume when blood pressure falls. *Glucocorticoids* (cortisol) are produced in response to ACTH from the anterior pituitary during stress to mobilize energy reserves. Cortisol (hydrocortisone) raises blood glucose levels and is anti-inflammatory. Small amounts of androgens are also produced in the innermost layer of the adrenal cortex in response to ACTH and other factors.

> Adrenal cortex is divided into three layers that secrete distinct hormones:
> mineralocorticoids (aldosterone), glucocorticoids (cortisol), and androgens
> Adrenal medulla secretes epinephrine and low levels of norepinephrine

The series of reactive responses initiated by hypothalamic CRF that leads to cortisol secretion is the HPA (hypothalamus-anterior pituitary gland-adrenal cortex) axis. The *HPA axis* is activated in response to stress to mobilize energy stores. As for secretion of thyroid hormones, cortisol feeds back to inhibit secretion of the tropic hormones CRF and ACTH.

The *hypothalamus-pituitary-adrenal (HPA) axis* is a good example of tropic hormone regulation. The hypothalamus produces CRH, corticotropin releasing hormone. CRH travels through the hypophyseal portal circulation of the infundibulum to the anterior pituitary gland where it stimulates release of the tropic hormone ACTH (adrenocorti-cotropin hormone) into the general circulation. ACTH acts on the adrenal glands, which sit atop the kidneys. Receptors for ACTH are in the outer layer (cortex) of the adrenal glands whose cells release corti-sol in response. Cortisol is the active hormone that increases blood sugar, suppresses the immune system and increases metabolic activity.

> Hypothalamus produces CRH (releasing hormone) that stimulates release of ACTH (tropic hormone) from anterior pituitary which causes production of cortisol (hormone) by adrenal cortex

Low cortisol levels in *Addison's disease* cause fatigue, weakness, and difficulty standing. The most common cause in the developed world is autoimmune disease. Tuberculosis was the most common cause when Addison first identified the disease. Both destroy the adrenal cortex. Cortisol deficiency is treated with replacement therapy (oral hydrocor-tisone).

Cushing's syndrome is due to elevated cortisol levels and causes weight gain and a characteristic moon face. Elevated cortisol may be due to long-term oral corticosteroid use (exogenous) or a tumor that causes overproduction of cortisol. *Cushing's disease* is a cause of Cushing's syndrome in which excess ACTH due to an adenoma or ele-vated CRH release overstimulates cortisol production by the adrenal cortex. Cushing's syndrome is treated by tapering off corticosteroids or by resection of the tumor, which then requires oral corticosterone re-placement therapy.

Other Hormones

Other important hormones are treated in the lectures on renal physiol-ogy (regulation of calcium levels by parathyroid hormones and calcitonin), digestion (blood glucose control by insulin and glucagon) central nervous system (circadian rhythms and melatonin), and repro-duction (LH, FSH, sex steroids).

Review Questions

1. How does communication in the endocrine system differ from that in the nervous system?
2. If hormones circulate throughout the body, how can they be specific?
3. What kinds of activities do hormones regulate?

4. How does the hypothalamus causes secretion of hormones from the anterior and posterior pituitary glands?

5. Make a table of the hypothalamic releasing hormones, the tropic hormones secreted by the anterior pituitary glands, the glands they stimulate, and the hormones each gland secretes.

6. How are thyroid hormones synthesized and released?

7. What are the primary effects of thyroid hormones on the body?

8. What are the two most common causes of thyroid deficiency? Why does iodine deficiency cause goiter?

9. What is the mechanism of Grave's disease?

10. How does growth hormone exert its effects?

11. Why do the effects of growth hormone hypersecretion prior to puberty differ from its hypersecretion in adults?

12. What are the three classes of hormones secreted by the adrenal cortex? Which is the primary hormone secreted in response to ACTH?

13. What are the causes and effects of Addison's disease and Cushing's syndrome?

Autonomic Nervous System

The second control system that regulates body organ function is the *autonomic nervous system* (ANS). Like the endocrine system, it acts through binding of chemical messengers to cellular receptors. Since the ANS is part of the nervous system, it uses *neurons* to send signals to body organs and glands. Every neuron has an *axon* that rapidly carries electrical signals (action potentials) long distances to specific sites. The ends of the axons (*axon terminals*) secrete *neurotransmitters* that act on body organs and glands to initiate a response that maintains homeostasis.

The ANS has two divisions In the sympathetic division, neurons exit the central nervous system from thoracic or lumbar regions of the spinal cord. In the parasympathetic, neurons exit directly from the brainstem (cranial nerves) or sacral regions of the spinal cord. Nearly all organs are innervated by both divisions. Sympathetic activity responds to threats. Parasympathetic activity dominates when the body is at rest. Neurons that exit the CNS are *preganglionic neurons* that synapse in *autonomic ganglia* with *ganglionic neurons* that innervate the end organ. Sympathetic activity also causes release of the circulating hormone epinephrine from the adrenal medulla, in the center of the adrenal glands that sit atop the kidneys. We finish by looking into *autonomic reflexes* and and the response to *physiological stress* as examples of ANS control.

Neurons

Describe the general features of a neuron

Neurons consist of a *cell body* (soma) where the nucleus is located and a single *axon* that carries electrical signals (*action potentials*) long distances to specific target cells. Most neurons have a set of dendrites that collect and bring signals into the cell body. Sensory neurons have sensory endings that directly couple to axons and a cell body off to the side of the neuron. Axons branch to end in a set of *axon terminals* where they release neurotransmitters. The site of communication between an axon terminal and another cell is a *synapse*.

Neurons have a cell body and single axon that carries action potentials to axon terminals where neurotransmitters carry signals across a synapse
Most neurons have dendrites that bring signals into the cell body (soma)

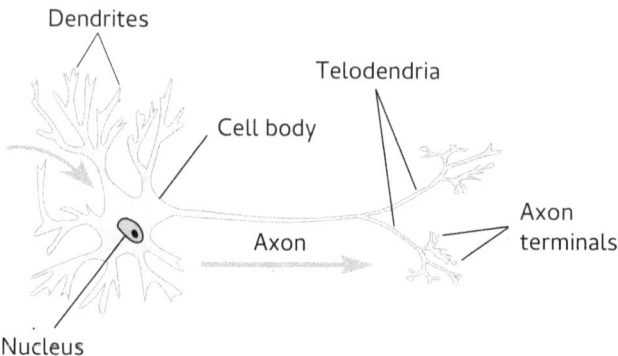

Autonomic Centers

List the brain centers that regulate autonomic functions

The *autonomic nervous system* (ANS) is that part of the peripheral nervous system that controls visceral (organ) functions. It is influenced by several centers in the brain. Most of these act through the *hypothalamus*, which integrates input and sends signals to visceral motor nuclei in brainstem and spinal cord. A nucleus is a site in the central nervous system (brain and spinal cord) where one neuron communicates with another through a synapse. From the autonomic nucleus, neurons are sent out to innervate body organs. The hypothalamus also acts on the endocrine system to initiate release of hormones from the *anterior pituitary gland* and epinephrine and norepinephrine from the *adrenal medulla*.

In addition to the hypothalamus, autonomic reflex centers in the brainstem regulate cardiovascular and respiratory functions. The primary regulatory center in the brainstem is the nucleus of the solitary tract (*solitary nucleus*), located centrally in the medulla oblongata. The *medulla oblongata* is the most inferior part of the brain, sitting directly atop the spinal cord.

Simple reflexes in the solitary nucleus respond to sensory input from *interoceptors* that report on internal body function, sending out signals to nuclei of the brainstem that project to preganglionic neurons. The solitary nucleus also sends signals to the hypothalamus, reticular formation (that is responsible for wakefulness), amygdala (the limbic fear center) and other brain centers. These centers help coordinate complex autonomic responses.

> Medulla oblongata has reflex centers for respiration and heart rate
> Hypothalamus controls viscera through activation of ANS and
> release of hormones by pituitary gland and adrenal medulla

Divisions of the ANS

Distinguish between the two divisions of the ANS

The ANS has two divisions: sympathetic and parasympathetic. The *sympathetic division* of the ANS passes out thoracic and lumbar spinal nerves, so is anatomically thoracolumbar. The *parasympathetic division* of the ANS surrounds the sympathetic division and exits from the CNS through cranial and sacral nerves (craniosacral outflow). One of the most important parasympathetic nerves is the *vagus nerve* (CN X), which wanders through the body to innervate major organs above and below the diaphragm.

> Sympathetic division exits CNS through thoracic and lumbar nerves
> Parasympathetic division exits through cranial and sacral nerves
> Most organs are innervated by both divisions of the ANS

The activities of the sympathetic nervous system increase heart rate and inhibit digestion. Sympathetic nervous activation is a response to threats, so it is called a "fight-or-flight" response. Activation of the parasympathetic division occurs when the body is at rest, which it should be after a meal. So, parasympathetic activities are called "rest-and-digest." They result in decreased heart rate and increased gastrointestinal activity. The activities of the sympathetic and parasympathetic systems are complementary: what one activates, the other generally inhibits. Thus, most organs are innervated by both divisions of the ANS (*dual innervation*) to provide tight homeostatic control.

Autonomic Ganglia

Describe the wiring of neurons through autonomic ganglia

Signals coming from higher autonomic centers in the brain (hypothalamus and brainstem) synapse with *preganglionic neurons* of the ANS. These neurons leave the CNS to synapse with (post)*ganglionic neurons* that innervate body organs. Thus, the output of motor neurons from the ANS passes through two neurons.

Preganglionic neurons arise in *autonomic nuclei* of the brainstem and lateral horn of the spinal cord. From the brainstem, preganglionic neurons leave the CNS through cranial nerves; those arising in spinal cord nuclei exit through spinal nerves. Preganglionic neurons synapse with (post)ganglionic neurons in peripheral *autonomic ganglia*. These *ganglionic neurons* directly innervate tissues, such as the myocardium of the heart and smooth muscle of the intestine.

> In autonomic nervous system (ANS) preganglionic neurons leave CNS
> to synapse in autonomic ganglia near spinal cord or innervated organs
> Autonomic ganglia send out ganglionic neurons that innervate tissues

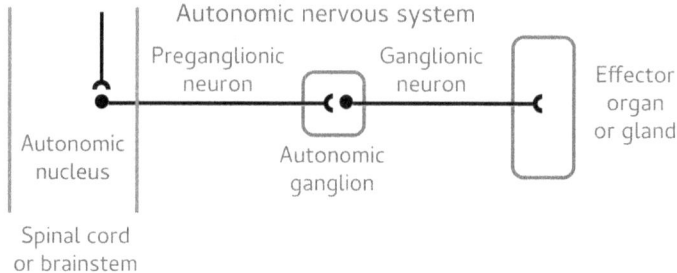

Autonomic ganglia in the parasympathetic division are located in or near the organs they innervate. Sympathetic nerves pass through or synapse in paravertebral ganglia of the *sympathetic chain*, next to the spinal cord. Extensions of the chain into the cervical and sacral regions distribute sympathetic neurons into those regions. Sympathetic neurons that innervate organs above the diaphragm synapse within the sympathetic chain.

> Sympathetic nerves from thoracic regions pass directly through autonomic
> ganglia of sympathetic chain to innervate organs above diaphragm

Sympathetic neurons that innervate organs below the diaphragm pass through the paravertebral ganglia without forming synapses. They then synapse in *collateral ganglia* in the abdominal cavity and near the pelvic organs they innervate.

> Sympathetic nerves from lumbar regions pass through sympathetic chain to
> collateral autonomic ganglia that innervate organs below the diaphragm

Parasympathetic ganglia are always near the organs they innervate. The parasympathetic *vagus nerve* (cranial nerve X) wanders through the viscera above and below the diaphragm. It supplies parasympathetic output to most body organs and receives sensory input from those same organs.

> Vagus nerve (cranial nerve X) supplies parasympathetic output
> to most organs above and below the diaphragm

Varicosities

Describe the structure and role of varicosities

In the somatic nervous system that innervates skeletal muscles, a single motor neuron sends signals to a limited number of muscle fibers, which constitutes a motor unit. If a stronger contraction is required, more motor neurons fire causing more motor fibers to be activated. The synapse is highly specific: each axon terminal forms a neuromuscular junction at one particular location on a single muscle fiber. This provides a great deal of control over how strongly a muscle contracts.

> Junctions of autonomic ganglionic neurons with target organs consist of swellings (varicosities) that bathe tissue with neurotransmitter

With viscera, rather large regions are activated all at once. It would not do to have just a part of the heart contract strongly while the rest of the heart contracts weakly. Instead of having an axon terminal go to one cardiac muscle cell, the ANS bathes the myocardium with neurotransmitter. Expanded regions along the axon terminus (*varicosities*) contain secretory vesicles that release neurotransmitter as the action potential passes through. The result is a broad dispersion of the neurotransmitters and activation of the entire tissue.

Adrenal Medulla

Describe the sympathetic activation of the adrenal medulla

The adrenal glands lie atop the kidneys. The outer part is the adrenal cortex, which secretes cortisol, in addition to aldosterone and androgens. The inner part is the *adrenal medulla* which is completely surrounded by the adrenal cortex. Secretions of the adrenal medulla increase in stressful situations. Sympathetic axons of preganglionic neurons pass through the celiac (collateral) ganglia in the upper abdomen, near the adrenal gland to terminate within the adrenal medulla.

> Preganglionic neurons pass through sympathetic chain and celiac ganglia to synapse within adrenal medulla

Adrenal medullary cells are modified postganglionic neurons that have no axons or dendrites (*chromaffin cells*). Preganglionic sympathetic neurons release acetylcholine from their axon terminals where they synapse with chromaffin cells. This acetylcholine binds to specific receptors on the chromaffin cells, causing them to release mostly epinephrine (and some norepinephrine). Epinephrine enters the blood to cause a systemic sympathetic response to threats.

> Sympathetic activation of adrenal medulla causes release of epinephrine
> Epinephrine enters circulation to cause systemic sympathetic activation

Autonomic Reflexes

Describe the components of an autonomic reflex

Receptors respond to stimuli. We have already considered several important receptors for hormones and neurotransmitters. These chemical messengers bind to protein receptors on the cell surface or inside the cell to cause a change in cell function. Another class of receptors are cellular receptors that respond to the condition of body fluids and organs. These receptors are specialized neurons: they carry electrical signals (action potentials) long distances, from the site of stimulus into the central nervous system (brain and spinal cord).

Sensory receptors are broadly divided into two types. *Exteroceptors* sense the external environment. These include sensory receptors in the skin, for touch, pressure, and pain and include the special senses of vision, hearing, balance, taste, and smell. *Interoceptors* sense the internal environment. These are the receptors that constantly monitor body functions.

> Autonomic reflexes respond to internal conditions of the body
> by sending correcting signals to organs and glands

An *autonomic reflex* consists of a receptor, control center, and effector that work together to maintain homeostasis. The receptors may be *thermoreceptors* that respond to temperature, c*hemoreceptors* that respond to chemicals, or *mechanoreceptors* that respond to stretch or pressure. Mechanoreceptors that respond to blood pressure are given the special name *baroreceptors*. These receptors send signals to the CNS (central nervous system) where the current condition of the body (such as blood pressure) is compared with its optimal value (setpoint). Correcting signals are sent to effectors that restore the controlled condition to its setpoint (negative feedback).

Pupillary Reflex

Describe the sympathetic response of the pupils to threats and the parasympathetic pupillary reflex to bright light

Autonomic reflexes are not always responses to changes in body organs and fluids. The pupil is the dark opening in the center of the iris. It regulates the amount of light that enters the eye. *Dilation* of the pupil is a sympathetic response to threats, due to contraction of pupillary dilator muscles. This contraction is mediated by norepinephrine binding to alpha-1 adrenergic receptors on the smooth muscle cells of the pupillary dilator muscle.

In bright light, pupil diameter gets smaller. This is a parasympathetic *pupillary reflex* that protects the retina from excess light. The sensory receptors are photosensitive retinal ganglion cells that are distinct from the photoreceptors (rods and cones) used for vision. The parasympathetic response arises from the Eddinger-Westphal nucleus in the midbrain. It passes along the oculomotor nerve (CN III) to the ciliary ganglion behind the eye, where it synapses with the postganglionic neuron that innervates the pupillary sphincter muscles. Released acetylcholine acts on muscarinic cholinergic receptors to cause muscle contraction and pupillary constriction.

> Pupillary dilator muscle causes pupils to dilate in sympathetic response
> Pupillary light reflex is parasympathetic response to bright light acting on photosensitive retinal ganglion cells that causes constriction of pupils

Sweat Glands

Explain how sympathetic and parasympathetic discharges alter sweat gland secretion

Adults have two types of sweat glands. *Ecrrine sweat glands* are responsible for producing sweat that cools the body as it evaporates. They are distributed over nearly the entire body, with highest density in palms and soles of feet. They secrete a watery fluid through sweat ducts that emerge on the body surface. Secretory cells in eccrine sweat glands are surrounded by myoepithelial cells. Sympathetic ganglionic neurons to sweat glands secrete acetylcholine that acts on muscarinic receptors to increase sweat production.

Apocrine sweat glands secrete onto hair shafts in pubic and axillary regions. They begin to function at puberty and are not as important for temperature control in humans. Apocrine sweat glands secrete an oily sweat in response to stress and sexual arousal (circulating epinephrine). They are a site of bacterial growth that causes body odor.

> Eccrine sweat glands secrete watery fluid that evaporates to cool body
> Sweat secretion is under sympathetic control via muscarinic receptors

Stress Response

Describe the phases of the response to stressors

Physiological stress is the response of the body to a threat (stressor). It initially causes a *sympathetic* response. Epinephrine is secreted by the adrenal medulla in response to hypothalamic innervation and norepinephrine is secreted by the sympathetic division of the ANS (sympathoadrenal medullary axis). This *alarm stage* increases heart rate, cardiac output, and blood pressure, and diverts blood flow to the heart and skeletal muscles in order to overcome or flee from the threat.

If the stressor is not removed, the body enters the *resistance stage*. To maintain a state of alert, long-term metabolic adjustments are made for increased energy needs. This resistance stage is driven by glucocorticoids (mostly cortisol) secreted by the adrenal cortex in response to ACTH produced in the anterior pituitary gland which has been stimulated by CRH from the hypothalamus (hypothalamic pituitary axis).

Long-term stress places a severe strain on the body. In its final phase, the *exhaustion phase*, vital systems collapse due to the failure of the body to meet its continuing need for energy. These three stages of the response of the body to stress (acute, resistance, and exhaustion) is the General Adaptation Syndrome that was first proposed by Hans Selye.

> Stress response initially activates sympathoadrenal medullary axis
> causing secretion of epinephrine and sympathetic activation
> Later, hypothalamic pituitary axis (HPA) is activated with cortisol release

Review Questions

1. What are the brain centers that regulate autonomic functions?
2. Describe the anatomical organization of the ANS and the function of its divisions.
3. What is an autonomic ganglion? Distinguish between preganglionic and ganglionic neurons.
4. Distinguish between ganglia of the sympathetic chain and collateral ganglia.
5. What is the vagus nerve and what organs does it innervate?
6. What are varicosities and how are they important in the ANS?
7. How is the adrenal medulla activated by the sympathetic nervous system?
8. What are chromaffin cells and how do they release epinephrine?
9. What are the components of an autonomic reflex?

10. What causes pupils to dilate? What is the pupillary reflex?
11. What stimulates secretion of sweat?
12. What are the three phases of the stress response? What hormones are responsible for each phase?

ANS Neurotransmitters

Now that we have seen how the endocrine system and ANS regulate body activities, we turn to explain how ANS neurons communicate in the autonomic ganglia and with the glands and organs it innervates. Recall that neurons send action potentials (electrical signals) along axons to specific sites where neurotransmitters are released by axon terminals. The site where the axon terminal communicates with another cell is a synapse. Neurotransmitters pass across a synaptic gap to act on receptors in the membranes of postsynaptic cells. The effects may either to open ion channels (ionotropic) or pass signals into the cell to alter its activity (metabotropic). In the ANS, the neurotransmitters are almost always either acetylcholine or norepinephrine.

Acetylcholine passes signals through autonomic ganglia and is the most common neurotransmitter in ganglionic neurons of the parasympathetic division of the ANS. *Norepinephrine* is the neurotransmitter most often used by sympathetic ganglionic neurons. Their effects on glands and organs depends on the neurotransmitter and receptor subtype. Receptors that respond to acetylcholine are *cholinergic*. Cholinergic receptors in autonomic ganglia are always excitatory (they always pass the signal) and respond to nicotine, so are *nicotinic*. Cholinergic receptors in end organs also respond to muscarine, so are classified as *muscarinic*. Receptors that respond to norepinephrine (and circulating epinephrine) are *adrenergic*. Adrenergic receptors are classified as either *alpha* or *beta*.

Synapses

Describe the structure and features of synapses

Axons carry action potentials from the cell body of a neuron to a set of axon terminals that synapse with specific target cells. At synapses *axon terminals* release neurotransmitters. A *presynaptic neuron* is the neuron whose axon carries signals to the *synapse*. The *postsynaptic cell* receives signals at the synapse.

Synaptic vesicles in the axon terminals (synaptic boutons) contain neurotransmitters that are released by exocytosis. The *synaptic cleft* is the space between the axon of the presynaptic neuron and the postsynaptic cell. Neurotransmitters in the synaptic cleft bind to cell surface receptors. Most are then degraded by hydrolytic enzymes or taken back into the axon terminals or nearby astrocytes (specialized cells in the brain) to keep synaptic levels of neurotransmitter from rising too high.

> Neurotransmitters are released from synaptic vesicles at axon terminals to cross synaptic cleft and bind to receptors on postsynaptic cell
> Neurotransmitters are degraded by enzymes in synaptic cleft

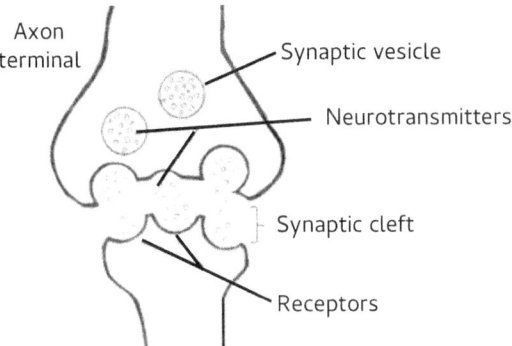

Neurotransmitters bind to cell surface receptors on the postsynaptic cell, or more rarely to receptors in the axon terminal of the presynaptic neuron. Neurotransmitter receptors are divided into two broad classes. One class directly opens membrane channels that allow ions to enter the cell. These are *ionotropic receptors*. The second is similar to the receptors for hydrophilic hormones: binding causes a change in cell function (the metabolic state of the cell). These are *metabotropic receptors*.

Ionotropic Receptors

Describe the location and functions of ionotropic receptors

The neurotransmitter in autonomic ganglia is acetylcholine. Binding to the cholinergic receptor in the ganglion is always excitatory, so signals are always passed through from preganglionic to ganglionic neurons. The receptors are also activated by the drug nicotine, so they are called *nicotinic cholinergic receptors*. Nicotinic cholinergic receptors are ionotropic. Binding of acetylcholine opens an ion channel in the cell membrane that allows ions to pass through and make it more likely that an action potential will fire in the postsynaptic cell.

Ionotropic receptors are *ligand-gated ion channels*. They respond to binding of a ligand by changing conformation (protein structure) from a closed state to an open state. In the *closed state*, ions cannot pass across the cell membrane through the channel. In the *open state*, ions that are able to pass through the channel do so. They may enter or leave the cell, depending on the concentration of ions inside and outside the cell and the charge on the membrane. Ligand-gated channels are found throughout the nervous system and include receptors for acetylcholine, serotonin, glycine, and glutamate.

Which ions flow through the membrane depends on the ion channel. For example, when the nicotinic cholinergic receptor opens in response to binding of acetylcholine, both sodium and potassium ions can pass through the channel. However, positively charged potassium ions (K^+) are kept inside cells by the negative charge on the inner surface of the plasma membrane. Sodium ions readily (Na^+) pass through the open channel down their concentration gradient and toward the negative internal charge.

> Ligand-gated channels open when ligand binds to allow passage of ions
> Nicotinic cholinergic receptor binds acetylcholine to allow Na ion entry

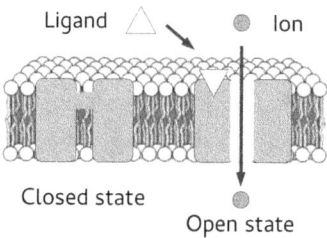

Nicotinic cholinergic receptors (nAChR) mediate transmission of signals between presynaptic and postsynaptic neurons in autonomic ganglia of the ANS, in the adrenal medulla, between motor neurons and skeletal muscle fibers, and in many sites within the brain. When acetylcholine binds to this receptor, a cation channel opens to allow Na ions to enter the postsynaptic cell. It is a cholinergic receptor because it responds to binding of the ligand acetylcholine. It is nicotinic because it also responds to the tobacco plant alkaloid nicotine.

Neurotransmitter Recycling

Explain how neurotransmitters are recycled at synapses

Acetylcholine is synthesized with the axon terminals of cholinergic neurons. It is packaged into synaptic vesicles which fuse with the axonal membrane at synapses, releasing the neurotransmitter into the synaptic cleft. After binding and being released from it postsynaptic receptor, acetylcholine is hydrolyzed by *acetylcholinesterase* that is located on the postsynaptic membrane. This inactivation prevents neurotransmitter levels from rising too high in the vicinity of the receptor. Choline is recycled back into the axon terminals by a specific *choline transporter*. This choline is re-used for synthesis of new transmitter molecules. Norepinephrine is taken back into presynaptic axon terminals intact through the *norepinephrine transporter* (NET).

> Neurotransmitters produced in synaptic vesicles of axon terminals are secreted into synaptic clefts and taken up by norepinephrine transporter or choline transporter after hydrolysis by acetylcholinesterase

Metabotropic Receptors

Describe signaling by metabotropic receptors

Metabotropic receptors link binding of neurotransmitter to production of a small second messenger molecule inside the cell. Most metabotropic receptors are G-protein coupled (GPCRs): ligand binding activates or inhibits a small G protein on the inner cell membrane surface. The G protein may activate or inhibit adenylate cyclase to alter cAMP concentrations or activate phospholipase C, which produces the second messenger *inositol triphosphate*. These second messengers activate or inhibit cellular enzymes or ion channels, altering the metabolism of the cell.

Metabotropic receptors play a critical role in the nervous system. They include receptors for dopamine, serotonin, norepinephrine, and acetylcholine. Metabotropic acetylcholine receptors that respond to the mushroom poison muscarine as an agonist are *muscarinic cholinergic receptors*. Muscarinic receptors are classified into five subtypes. All five are important in the brain.

> G-protein coupled receptors acting through Gq activate phospholipase C (PLC) that produces IP3 second messenger activation of Ca release and DAG activation of protein kinase C

In the peripheral nervous system, muscarinic cholinergic receptors mediate the "rest-and-digest" functions of the parasympathetic division of the autonomic nervous system (ANS). M1 and M3 receptors are coupled with G_q proteins that stimulate phosphoinositide hydrolysis and the release of intracellular calcium. In some tissues, they also activate the signaling enzyme protein kinase C. These subtypes increase secretions by salivary glands and the stomach. They also mediate sympathetic activation of sweat gland secretions. M2 is coupled with G_i proteins (inhibitory) that reduce cAMP production and open K ion channels in cardiac pacemakers to slow heart rate.

Adrenergic Receptors

Describe the subtypes of adrenergic receptors

Adrenergic receptors are metabotropic receptors (GPCRs) that respond to binding of the catecholamines epinephrine and norepinephrine. The name comes from the British use of adrenaline for epinephrine. They are the primary receptors that mediate the "fight-or-flight" response of the sympathetic division of the autonomic nervous system. Norepinephrine is the primary catecholamine released by postganglionic neurons of the sympathetic nervous system. Epinephrine is the primary circulating catecholamine, which is secreted by the adrenal medulla. Adrenergic receptors are divided into two groups: α (*alpha*) and β (*beta*). Each group has multiple subtypes, which act through distinct GPCR-mediated mechanisms that mediate sympathetic responses.

ANS Neurotransmitters

Define the neurotransmitters used in the autonomic nervous system at autonomic ganglia and target organs

In the autonomic ganglia, signals always pass through from preganglionic to postganglionic neurons. The neurotransmitter is *acetylcholine* and the effect is always excitatory. This is due to direct opening of ion channels when acetylcholine binds to *nicotinic cholinergic receptors*. Such receptors are *ionotropic*. They allow Na ions to flow into the cell, causing local depolarization (decreased negative charge on the inner membrane surface) when a ligand binds. Spread of depolarization to the axon of the postganglionic neuron initiates an action potential that travels the length of the axon to the tissue that is innervated.

At the tissues, where ganglionic neurons synapse, the neurotransmitters for the sympathetic and parasympathetic divisions of the ANS differ. The primary neurotransmitter in sympathetic ganglionic neurons is *norepinephrine*. Sympathetic activation also causes release of epinephrine from the *adrenal medulla*. The British name for norepinephrine is noradrenaline and that for epinephrine is adrenaline. Thus, receptors in sympathetic postganglionic neurons are called adrenergic *receptors*. Adrenergic receptors are coupled to intracellular G-proteins that lead to a change in cell function.

> Synapse of preganglionic with ganglionic neuron is nicotinic cholinergic
> Parasympathetic ganglionic neuron is muscarinic cholinergic
> Sympathetic ganglionic neuron is adrenergic

The primary neurotransmitter in the parasympathetic division of the ANS is *acetylcholine*, just as it is in the autonomic ganglia. However, instead of the excitatory ionotropic receptors found in autonomic ganglia, the receptors in tissues innervated by the parasympathetic division are *metabotropic*, as are the adrenergic receptors. Since they also respond to the drug muscarine, they are called *muscarinic cholinergic receptors* In both cases, binding of neurotransmitter causes a response inside the cell that alters the activity of the tissue.

Neurotransmitter Release in Adrenal Medulla

Explain the mechanism of neurotransmitter release in adrenal medulla

Sympathetic neurons enter the adrenal medulla from the celiac ganglia of the upper abdomen. There they release acetylcholine which acts on nicotinic cholinergic receptors of the *chromaffin cells* in the adrenal medulla to cause release of epinephrine. Epinephrine passes through the interstitial fluids to enter the blood and act as a hormone that causes systemic sympathetic activation.

The release mechanism is the same as that in axon terminals of neuronal synapses throughout the body. Arrival of an action potential at the axon terminal of the preganglionic neuron opens Ca ion channels. Ca ions enter the cell to cause fusion of synaptic vesicles with the axonal membrane, releasing acetylcholine into the synaptic cleft. Binding of acetylcholine to nAChR of the chromaffin cell opens Ca ion channels that cause exocytosis of epinephrine.

> Release of epinephrine from chromaffin cells of adrenal medulla is activated by Ca entry through voltage-gated Ca channels that open in response to acetylcholine binding to nicotinic cholinergic receptors

Adrenergic Receptor Subtypes

Distinguish the adrenergic receptor subtypes alpha-1, beta-2, and beta-2

Blood vessels in skin, mucosa, and abdominal viscera have *alpha-1* (α-1) *receptors* that respond to norepinephrine released from axon terminals of sympathetic ganglionic neurons. These GPCR receptors are coupled to G_q, so activate phospholipase C which leads to release of calcium and smooth muscle cell contraction. Contraction of smooth muscle cells surrounding small arteries (arterioles) that lead to capillary beds causes vasoconstriction (a decrease in vessel diameter). This reduces blood flow to the skin and gastrointestinal (GI) tract and increases systemic blood pressure by making it harder to pump blood through the extensive blood vessels of these body regions.

The *alpha-2* (α-2) *receptor* is located presynaptically, in the membrane of the axon terminal of neurons in the brain. Release of norepinephrine feeds back to the presynaptic neuron by binding to its α-2 receptor which is coupled to G_i proteins. The inhibitory effect of G_i decreases norepinephrine release and reduces signal transfer to postsynaptic neurons.

> Alpha-1 adrenergic receptors bind NE to cause vasoconstriction
> Beta-1 receptors bind E and NE to increase heart rate and contractility
> Beta-2 receptors bind epinephrine to cause vasodilation and bronchodilation

All β-receptors are coupled to G_s which increases cAMP levels. Most *beta-1* (β-1) *receptors* are located in the heart. They respond to both circulating epinephrine and to norepinephrine released by sympathetic neurons. The effects of ligand binding are to increase heart rate (*chronotropic*) , velocity of signal conduction through the myocardium (*dromotropic*), and strength of cardiac muscle contraction (*inotropic*).

Beta-2 (β-2) receptors are more sensitive to epinephrine than norepinephrine. They cause smooth muscle cell relaxation in bronchi (bronchodilation) and the GI tract. In active skeletal muscles during exercise, vasodilation mediated by β-2 receptors dominates over the usual sympathetic vasoconstriction caused by α-1 receptors.

ANS Drugs

Describe the mechanisms by which drugs acting on ANS receptors alter body function

Drugs and poisons can be used to tease out mechanisms. Consider an inhibitor of acetylcholinesterase, such as the organophosphate insecticide *parathion*. Parathion slows the breakdown of acetylcholine in the synaptic cleft, leading to increased ACh concentrations. The expectation is that any synapse that uses ACh as neurotransmitter will have increased activity. Parathion has both nicotinic and muscarinic effects. The muscarinic effects are parasympathetic: salivation, bronchoconstriction, and bradycardia (lowered heart rate), all of which can be life-threatening. *Atropine* is an antagonist at muscarinic cholinergic receptors that reverses the adverse muscarinic effects of parathion.

Like atropine, *ipratropium* blocks muscarinic receptors. Its primary use is as a bronchodilator in COPD (chronic obstructive pulmonary disease). Administration by inhalation ensures that most of the drug acts on the bronchi while minimizing its systemic side effects.

Nicotine is an agonist of nicotinic cholinergic receptors. At autonomic ganglia, it activates both divisions of the ANS. In the CNS, it increases release of dopamine and norepinephrine, neurotransmitters associated with pleasure and weight control.

Bethanechol is an acetylcholine analog that acts on muscarinic cholinergic receptors. Thus, it has parasympathetic effects: it increases gut motility, activates the detrussor muscle of urinary bladder, and relaxes its trigone and sphincter muscles. This drug is primarily used in nonobstructive urinary retention to stimulate an atonic bladder postpartum or after surgery. Its adverse effects are consistent with muscarinic activation: sweating, salivation, flushing, decreased blood pressure, diarrhea, and bronchospasm.

Epinephrine is a natural sympathetic hormone that also has therapeutic uses. Epinephrine activates beta-2 adrenergic receptors to cause bronchodilation. Thus, it is the primary drug used in emergency treatment of bronchoconstriction associated with anaphylactic shock. Epinephrine also activates beta-1 receptors in the heart, increasing heart rate and strength of contraction. For this reason, it can be used to restore cardiac rhythm after cardiac arrest. *Norepinephrine* primarily acts on alpha-1 receptors, causing vasoconstriction and increased blood pressure. It is used to treat cardiovascular shock (sudden loss of blood pressure).

Doxazosin is a selective, competitive blocker of alpha-1 adrenergic receptors which are responsible for vasoconstriction. It lowers blood pressure by blocking sympathetic signals to arteries and veins, resulting in vasodilation and decreased resistance to blood flow.

Propranolol is a nonspecific beta blocker, having equal effects on beta-1 and beta-2 receptors. Blocking beta-1 receptors decreases heart rate and contractility (strength of contraction) which reduces cardiac output and reduces the work of the heart, relieving the pain of stable angina. Propranolol is also used to treat hypertension and prophylactically for migraine. Blocking beta-2 receptors may cause bronchoconstriction leading to a respiratory crisis in patients with COPD, so the drug is contraindicated in such patients. Cardioselective beta blockers that act specifically on the cardiac beta-1 adrenergic receptor, such as *metoprolol* may be used to treat angina and hypertension in patients with impaired pulmonary function.

Review Questions

1. How do neurotransmitters differ from hormones?

2. Why are neurotransmitters either degraded or taken up by presynaptic or surrounding cells?

3. What is a ligand-gated ion channel? Why are they also called ionotropic receptors? Give an example.

4. Where are nicotinic cholinergic receptors found in the body?

5. Describe the stages of acetylcholine signaling and recycling at a nicotinic cholinergic receptor.

6. How do metabotropic receptors differ from ionotropic receptors? Give an example of a metabotropic receptor.

7. What is the mechanism of release of epinephrine from adrenal medulla?

8. How do cellular responses to alpha-1, beta-1, and beta-2 adrenergic receptors differ?

9. How does signal transduction by G_S, G_I, and G_Q proteins differ?

10. What neurotransmitters and receptors are most common in the neurons of the ANS?

11. How can drugs be used to better understand molecular mechanisms in the body?

Excitable Cells

Given the rules of ionic electricity, the major biological problem in understanding action potentials is to describe and explain the ion permeability mechanisms in the membrane.

– Bertil Hille

Membrane Transport

Body functions are regulated by binding of molecules to receptors. Chemical messengers may be hormones that circulate throughout the body to act on target cells or neurotransmitters that are released from synapses at their sites of action. The response of a cell to a chemical messenger depends on the messenger molecule and its specific receptor. Often, this response is contraction of muscle tissue or initiation of an electrical signal in a postsynaptic cell. The upcoming set of lectures, beginning with this introduction to membrane transport, will explain how muscles contract and nerves conduct signals. The primary notion to keep in mind is that nature tends toward equilibrium: molecules diffuse to equalize concentrations and minimize charge differences.

We begin this lecture with an explanation of molecules get across cell membranes. Permeable molecules, such as steroid hormones and water, readily pass through membranes by *simple diffusion*. Water-soluble molecules and ions require a carrier protein to pass through a membrane. This is *facilitated diffusion*. Both types of diffusion are always from high concentrations to low concentrations. The transport of water across a membrane is given the special name *osmosis*. To go the other direction requires energy so is called *active transport*. Sodium ions are actively transported out of cells to maintain *osmotic balance* with the high protein concentrations inside cells. Potassium ions leak through membranes until a charge builds up on the inside of the cell membrane that opposes their flow. This is the negative *membrane potential*.

Membrane Permeability

Define diffusion and list the kinds of molecules that diffuse across membranes

Nature tends toward making all things equal. If this principle is applied to the concentration of molecules in a solution, we expect molecules to diffuse from a higher to lower concentration until the concentration is equal throughout the solution. The distribution of molecules will remain unchanged unless disturbed, a condition called *equilibrium*. The driving force that causes molecules to move from a higher to lower concentration is a *chemical gradient*.

Diffusion causes an equal distribution of molecules throughout a solution
Each molecule follows a random path that results in complete mixing

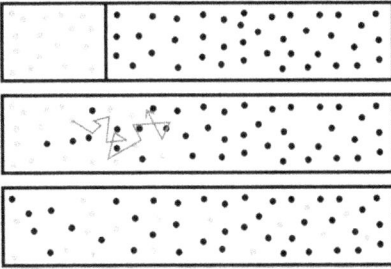

Solute particles in a solution mix to equalize their concentration throughout. This localized spreading of molecules is *diffusion*. Diffusion is the process that moves oxygen across the very narrow space between the air sacs of the lungs and the blood. The direction of diffusion is from higher to lower concentrations of solute. This direction is said to be down the concentration gradient. *Bulk transfer* is the movement of materials over long distances, such as the exchange of stale for fresh air in the lungs.

Permeable molecules (oxygen, steroids) pass across lipid bilayers
until their concentrations are the same on both sides

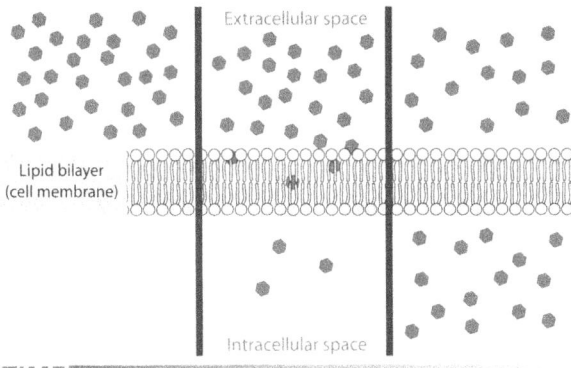

Diffusion is rapid over microscopic distances, but the time for diffusion increases exponentially with distance. Cells are never more than a few microns from a capillary, so diffusion of oxygen takes only a few msec in tissues. However, it takes more than 6 hours for oxygen to diffuse 1 cm. Getting oxygen from the lungs to other parts of the body requires a circulatory system.

The cell membrane is *semipermeable*. Not all molecules can pass through the membrane. If a molecule is permeable, it will diffuse across the membrane. Large polar molecules and ions do not pass across the lipid bilayer of the cell membrane. Water and small, nonpolar molecules (including gases such as oxygen and carbon dioxide) pass through the plasma membrane by *simple diffusion*. Simple diffusion does not require energy or proteins. The direction of diffusion is the same as in solution, from high to low concentrations of solutes.

Osmosis

Explain the process of osmosis and what determines its direction

The passive transfer of water across a cell membrane is *osmosis*. Water flows toward solutions with high particle concentrations in an attempt to dilute them. Only the number of particles matters and not their size or what kind of molecules they might be, so long as they cannot diffuse back across the membrane. *Osmolarity* is the molar concentration of impermeant particles. Water moves across a semipermeable membrane toward the solution with higher osmolarity.

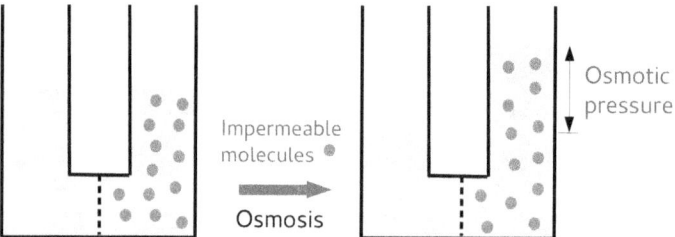

Osmosis is transfer of water across a semipermeable membrane
from a lower solute concentration to a higher one
Osmotic pressure is pressure required to stop the flow of water

Osmotic pressure measures the drawing power of water as it is pulled across the membrane toward the more concentrated solution. As water transfers across, a pressure is built up in the more concentrated solution that opposes the flow. *Osmotic pressure* is the pressure which needs to be applied to stop the flow of water. If two solutions are separated by a membrane that is not permeable to the solute, water flows toward the more concentrated solution by osmosis. Pressure builds up due to the increased height of the column of water. At equilibrium, the gravitational force per unit area that results is osmotic pressure.

Tonicity

Describe the relationship between solution osmolarity and cell volume

Cells must maintain *osmotic balance* with their surroundings so they do not shrink or burst. *Tonicity* determines whether a cell shrinks or swells when placed in a solution. *Isotonic* solutions have no effect on cell volume. They maintain osmotic balance. *Hypertonic* solutions cause cells to shrink; water leaves the cell. *Hypotonic* solutions cause cells to swell; water enters the cell. Shrinkage of red blood cells in hypertonic solutions is crenation. Swelling and bursting of red blood cells in hypotonic solutions is hemolysis.

> Extracellular fluids must remain isotonic to intracellular fluids
> Cells shrink in hypertonic (concentrated) solutions
> Cells swell in hypotonic (dilute) solutions

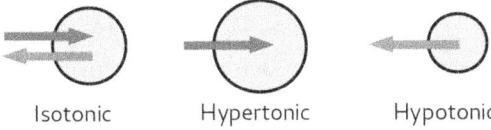

Isotonic   Hypertonic   Hypotonic

Facilitated Diffusion

Explain the process and give examples of facilitated diffusion

Molecules that cannot pass directly through membranes are aided by membrane proteins that act as channels or carriers. This membrane protein-mediated transport is *facilitated diffusion*. The direction of facilitated diffusion is from higher to lower concentrations, so does not require cellular energy.

> Molecules that cannot spontaneously pass through a
> lipid bilayer (glucose, amino acids, ions) require membrane proteins
> to assist transport by facilitated diffusion

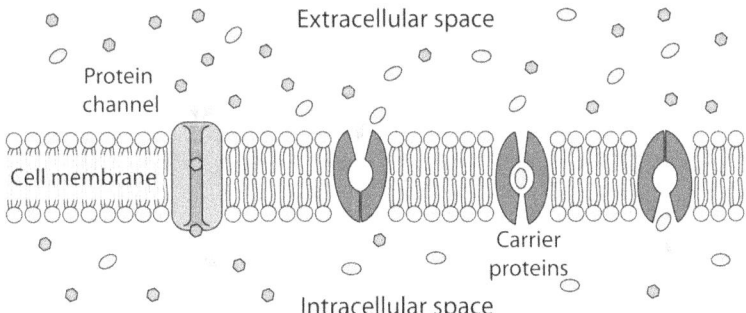

Carrier proteins bind specific molecules and are limited in their transport capacity. These include important *transporters* for glucose and amino acids. Carrier proteins have been distinguished from channels by their limited capacity to transport molecules. If the kidney glucose transporter becomes saturated because plasma glucose levels are too high in an uncontrolled diabetic, glucose spills into the urine. *Ion*

channels are integral membrane proteins that allow ions to pass through membranes.

Gated Ion Channels

Explain how ion channels regulate whether ions pass through them

Ion channels form pores that allow ions to pass through cell membranes. Channels can be *gated* (open or closed) to control passage of ions across membranes. Many ion channels are gated, to allow ions to pass through only if a ligand is bound, the channel is deformed by pressure, or the membrane voltage changes. *Mechanically-gated* ion channels open in response to physical distortion of membrane. *Ligand-gated* ion channels (sometimes called chemically-gated channels) respond to binding of specific molecules. *Voltage-gated* ion channels respond to changes in voltage (membrane potential).

> Ion channels form pores that allow ions to enter cells
> Gated ion channels open when stimulated by ligands (chemicals),
> voltage changes or mechanical deformation

Active Transport

Distinguish facilitated diffusion from active transport

Movement of a solute across a membrane against its concentration gradient is *active transport*. Active transport requires energy and a carrier protein. It is commonly used to pump ions against a concentration gradient. Na ions are maintained at high concentrations outside cells by Na-K ATPase. Pumping Na ions out of cells is against its concentration and charge gradients, so requires energy. The energy for ion transport by the Na-K ATPase comes from cleavage of ATP to ADP.

> Facilitated diffusion: membrane proteins allow diffusion of molecules
> Active transport uses energy to move molecules up a concentration gradient

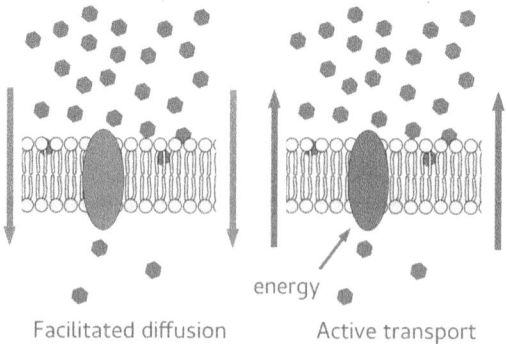

Facilitated diffusion   Active transport

Osmotic Balance

Describe how osmotic balance is maintained in cells

Inside cells, high protein and metabolite concentrations contribute to a high intracellular osmolarity. Outside cells, in the interstitial fluids, protein concentrations are much lower. Without a balancing of particle concentrations, water will slowly enter cells and cause them to burst. For this reason, cells pump Na ions out into the extracellular space so that the impermeable particle concentrations are the same on both sides of the cell membrane. This results in a high concentration of Na ions in extracellular fluids and a low concentration of Na ions inside cells.

> Na ions are pumped out of mammalian cells to balance high intracellular protein concentrations and maintain osmotic balance

Membranes are nearly impermeable to Na ions, so Na returns back into cells very slowly (down its concentration gradient). The Na pump (Na-K ATPase) maintains high extracellular Na ion concentrations by constantly pumping out most of the Na ions that re-enter. Thus, the high concentration of organic solutes inside the cell is balanced by a high concentration of Na ions outside the cells. On the other hand, cell membranes are relatively permeable to K ions because they have open (leak) channels that allow them to pass through the membrane bilayer. Unlike Na ions that are pumped out, K ions reach equilibrium across the membrane.

> Na-K ATPase uses energy from ATP hydrolysis to pump Na ions out of cells to maintain high extracellular Na ion concentrations and osmotic balance

Ouabain is a cardiac glycoside that inhibits the Na pump. At low doses, it slows the efflux of Ca ions out of cardiac muscle cells which increases their contractile force. It has been used to treat heart failure, but is no longer approved in the USA because of its high toxicity. Treatment of cells with ouabain causes them to swell, since Na can no longer be pumped out.

Resting Membrane Potential

Explain the origin of the resting membrane potential

Uncharged molecules diffuse from higher to lower concentrations across membranes, if the membrane is permeable to those molecules. Diffusion of charged molecules is also influenced by their charge. Just as molecules diffuse to equalize concentrations on both sides of the membrane, ions diffuse in a direction that equalizes charge. This combined gradient of concentration and charge is the *electrochemical gradient*.

Diffusion of ions across a membrane requires an open ion channel. Membrane ion channels that open randomly are *leak channels*. Leak channels for K ions are far more abundant than those for Na ions. Thus, K ions pass more readily across membranes than do Na ions. They have higher permeability.

What determines the distribution of these highly permeable K ions? The direction of diffusion of ions is determined by a balance of two forces: 1) molecules diffuse in a direction from high to low concentrations; 2) ions diffuse in a direction that reduces differences in charge (cations diffuse toward a negative charge). The high intracellular concentration of K ions drives them to diffuse out of the cell. However, K ions are also charged. For every K ion that diffuses out of the cell, a positive charge is brought along with it. As positively charged K ions leave the cell, the interior of the cell becomes slightly negative.

K ions leave cells through open channels down their concentration gradient
As K ions diffuse out, negative membrane potential opposes their flow

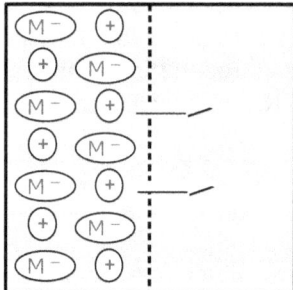 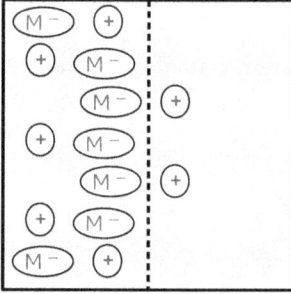

A balance is reached where the flow of K ions out of the cell (down their concentration gradient) is balanced by the negative charge that keeps them inside the cell. Since it is a combination of charge and concentration effects, the process is *electrodiffusion*. K ions diffuse across the membrane until a charge builds up opposing their flow. This negative membrane potential is the *resting membrane potential*. It is found in all cells at rest. The difference in electrical potential across the membrane is measured in units of volts (V). In the body, charge differences are very small, so voltage is measured in millivolts (mV), which is 1/1000 of a volt. The resting membrane potential is a tiny charge difference, on the order of -70 mV in most cells.

Membrane Potential and Ion Permeability

Describe the relationship between membrane potential and ion permeability

The relative permeability of ions and their distribution across the membrane determines the membrane potential. In neurons and muscle fibers, the Na pump maintains the Na concentration difference, while K ions distribute passively through leak channels. Permeability can rapidly change by opening and closing ion channels in the membrane. Very few ions pass across the membrane during an action potential or in response to a stimulus. The ion concentrations inside and outside of the cell remain nearly constant.

The magnitude of the membrane potential can be calculated from the concentrations of Na and K ions and their permeabilities using the *Goldman equation*. Since the concentrations of Na and K ions are maintained nearly constant by the Na pump, the magnitude of the membrane potential depends only on its permeability to the ions. If the membrane were permeable only to K ions, the predicted membrane potential (*equilibrium potential*) would be –90 mV. If the membrane was freely permeable to Na ions, the membrane potential would be +55 mV. Since neurons are much more permeable to K ions than to Na ions, the calculated membrane potential is –70 mV, which is a typical value for resting membrane potential.

> Concentrations of ions inside and outside cells remain nearly constant
> Value of membrane potential depends on relative ion permeability

Review Questions

1. What is the direction of diffusion of uncharged molecules, such as glucose?
2. What kinds of molecules can diffuse directly across cell membranes without requiring proteins?
3. What is osmosis? What determines its direction? What is meant by osmotic pressure?
4. What is the effect of a hypertonic solution on cells? Why would a hypotonic solution be a problem?
5. What types of molecules require a protein transporter or channel to pass through membranes? What is this process called?
6. What is a gated ion channel? What are the three classes of ion channels?
7. Why does active transport require energy?

8. How is osmotic balance maintained in cells?
9. Why do the K ions in a cell simply equilibrate across the membrane?
10. How are Na and K ions distributed between extracellular and intracellular fluids? How do their membrane permeabilities differ?
11. What happens if permeability of the membrane to Na ions increases?

Neurons

We have seen how many hormones bind to cell surface receptors that carry signals into cells while others enter cells to alter expression of DNA in the nucleus. Now we turn to signals that are carried along the membrane by *action potentials*. Action potentials are observed in the axons of neurons and in muscle fibers. They travel long distances without loss of signal strength. The key to this remarkable property are the voltage-gated Na and K channels found in their membranes. Once the first voltage-gated channel opens, it spreads an electrical signal that causes the channel next to it to open and then the next, all the way down the axon or muscle fiber.

Action potentials are initiated by local changes in membrane potential (graded potentials) that depolarize the cell enough to reach the *threshold* for opening of *voltage-gated Na channels*. Action potentials are of short duration in neurons because voltage-gated Na channels rapidly become *inactivated* after opening and are then unable to allow any more Na ions to enter the cell. At the same time, *voltage-gated K channels* open to repolarize the membrane. Until the Na channels reset to *closed*, they are unable to open and another action potential cannot fire.

At synapses, *voltage-gated Ca channels* open when action potentials arrive at the axon terminal. Influx of Ca ions from extracellular fluids into the cell causes *synaptic vesicles* containing *neurotransmitters* to release their contents into the *synaptic cleft* where they can bind to postsynaptic receptors. At the *neuromuscular junction* between motor neurons and skeletal muscle fibers, the receptor is a *nicotinic cholinergic receptor* that responds to *acetylcholine*. Binding of acetylcholine depolarizes the *motor end plate*. The resulting graded potential spreads nearby to the muscle fiber membrane where voltage-gated Na and K channels carry action potentials that travel the length of the fiber.

Classification of Neurons

Classify neurons according to their structure

Nerve tissues include *neurons* (excitable cells) and *neuroglia* (supporting cells). All neurons have a *cell body* (soma) with a nucleus and an axon that conducts signals. The single *axon* is a unique neuronal structure that passes electrical signals to other neurons, muscles, or glands. The site where an axon contacts the cells it innervates is the *synapse*. Synapses provide chemical communication to post-synaptic cells. The *axon hillock* is where the cell body joins the axon. The *initial segment* (near the axon hillock or sensory receptor) is where the action potential is initiated.

Neurons are classified structurally according to the location of the cell body relative to the axon. *Dendrites* carry impulses toward the cell body in many neurons. The most common neurons are *multipolar*: multiple dendrites enter the cell body and one axon leaves it. *Pseudounipolar* neurons have a single axon with a stem to the cell body that resides in a spinal ganglion. The peripheral axonal process of a pseudounipolar neuron is a *sensory ending* (not a dendrite) that carries signals from the periphery to the brain and spinal cord (CNS). *Bipolar* neurons have one dendrite that enters cell body and one axon that leaves it. Bipolar neurons are relatively rare; they are found in olfactory receptors and in the retina.

> Multipolar neuron: multiple dendrites enter cell body
> Pseudounipolar neuron: cell body is off to side of axon
> Bipolar neuron: single dendrite enters cell body

Changes in Membrane Potential

Describe how membrane potential can be changed

The *resting membrane potential* is a slight negative charge on the inside surface of the cell membrane of all cells. It is largely due to the relatively high permeability of membranes to K ions but not to Na ions and the activity of the Na-K ATPase that pumps Na ions out of cells. The key to understanding membrane potentials is to keep in mind that the distribution of ions in intracellular and extracellular spaces does not change. Outside the cell, the most abundant cation is Na ions. Inside the cell, it is K ions. What changes membrane potential in neurons and other excitable cells is a shift in permeability of the cell membrane to these ions.

In the resting state, the membrane is relatively permeable to K ions. As K ions diffuse out of the cell down their concentration gradient, a negative charge builds up on the inner membrane surface that keeps them from leaving. What happens if ion channels are opened for Na ions?

Na ions enter cells when the membrane is made permeable to them by opening of gated ion channels. The direction of movement is predicted by the *rules of ionic electricity:* diffusion from high to low concentrations is balanced by diffusion that minimizes charge differences. For Na ions, concentration and charge differences go in the same direction. The concentration of Na ions outside the cell is higher than inside, which drives Na ions into the cell. At the same time, the positively charged Na ions move toward the negative charge on the inner membrane surface.

> Opening of Na channels causes Na ion entry into cells in order to collapse both concentration and electrical gradients

As Na ion permeability increases and becomes much larger than the permeability of K ions, the membrane potential will approach +55 mV, This is the equilibrium potential for Na ions and is positive because of the high Na ion concentration outside the cell. High Na ion permeability only lasts a short time, so very few Na ions move across the membrane. The few that do so are returned to the extracellular fluids by the Na pump.

> Increased Na permeability depolarizes membrane potential (more positive)
> Return to normal low permeability repolarizes membrane to resting level
> Opening of K ion channels causes hyperpolarization (more negative)

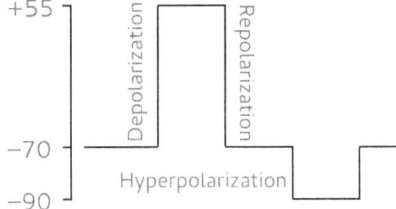

Thus, increases in Na ion permeability make membrane potential more positive. Since the resting membrane potential is rather far from zero, the membrane is said to be polarized. Increased Na ion permeability makes the membrane potential more positive, bringing it from about -70 mV toward zero. This is *depolarization*. When Na ion channels close and permeability decreases once again, the membrane potential returns to its resting membrane value. This is *repolarization*. Increased K ion permeability causes the membrane to become more negative. This is *hyperpolarization*.

Graded Potentials

Describe the origin and properties of graded potentials

Often, stimulation of a neuron or sensory receptor opens channels that allow Na ions to enter the cell. When a cation channel opens, Na diffuses down its charge and concentration gradients. Since the *concentration* of sodium is higher outside the cell than inside, the concentration gradient points toward the inside of the cell. The *charge gradient* for Na ions points in the same direction: Na ions are positively charged and the inner membrane surface is negatively charged. Thus, if the membrane is made permeable to Na ions by a stimulus, Na ions enter the cell. This entry of positive charges causes *depolarization*: the membrane potential becomes less negative.

The extent of depolarization of the membrane depends on the strength of the stimulus. A stronger stimulus opens more Na channels causing more depolarization. When the change in membrane potential depends on stimulus strength it is a *graded potential*. Graded potentials only spread out near the site where the stimulus occurs. Unlike action potentials, they cannot travel very far. Graded potentials are observed in post-synaptic cells at synapses, at sensory endings, and in dendrites and cell bodies of neurons.

> Opening of Na Ion channels allows Na ions to enter cells
> Increased Na ion permeability makes membrane potential more positive
> When channels close, low Na permeability returns potential to negative

Why does a graded potential depend on stimulus strength? Recall that membrane potential depends on the distribution of ions across the membrane and the permeability of those ions. The concentration of Na ions outside cells is high. Its concentration inside cells is low. But, at rest, the permeability of Na ions across the membrane is very small compared with that of K ions. Because of this low permeability, Na ions contribute very little to membrane potential. Resting membrane potential is largely due to the much higher permeability of K ions.

The concentration difference for ions across the membrane does not change. That is maintained by the Na pump. If the permeability of the membrane to Na ions increases, membrane potential becomes more positive. How much it changes depends on Na ion permeability. A larger increase in permeability results in a greater depolarization of the membrane. This increase in permeability to Na ions is directly related to the fraction of Na ion channels that are open. More open channels means greater permeability and more depolarization.

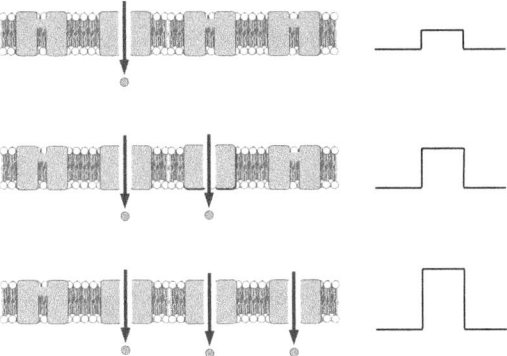

Na Ion permeability depends on number of open channels
Increased permeability results in greater depolarization

Voltage-Gated Ion Channels

Describe the properties and give examples of voltage-gated ion channels

Ligand-gated ion channels open in response to binding of chemicals to the receptor. Nicotinic cholinergic receptors are ion channels that open to allow Na ions to pass through acetylcholine binds. Mechanoreceptors respond to deformation, caused by direct pressure applied to the channel and its nearby membrane. *Voltage-gated ion channels* open and close in response to changes in membrane potential. The voltage at which a channels opens is its *threshold voltage*. Voltage-gated Na (Nav) channels in the initial segment of an axon opens when the membrane depolarizes to about -55 mV.

Three voltage-gated ion channels are significant: those for Na^+, K^+, and Ca^{2+} ions. Voltage-gated Na and K ion channels are found in axons and muscle fibers. They are responsible for the action potential that carries signals long distances along their cell membranes. Voltage-gated Ca ion channels are important at synapses and in the heart and smooth muscle cells.

Voltage-gated K ion channels can be either open or closed. The same applies to voltage-gated Ca ion channels. *Voltage-gated Na channels* have three states: open, closed, and inactivated. When *open*, Na ions pass through the channel down their electrochemical gradient. When closed or inactivated, no ions flow through the channel. What is the difference? When *closed*, the channel opens when membrane potential reaches threshold voltage. After staying open for a short time, the channels is *inactivated*. When inactivated, the channel must first close before it can open. This delays the re-opening of the channel and prevents it from opening again too quickly.

> Closed voltage-gated Na channels open at threshold (about -55 mV)
> Open channels allow ions to pass through and cause depolarization
> then inactivate and cannot re-open until they pass through closed state

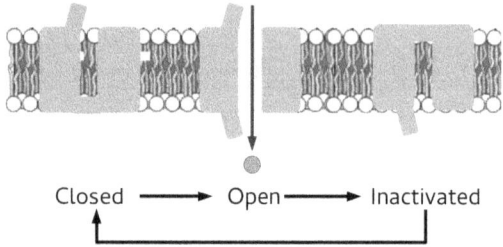

Closed ⟶ Open ⟶ Inactivated

Action Potentials

Explain the origin and stages of the action potential

Axons of neurons and muscle fibers transmit *action potentials* (electrical signals) along their membranes. An action potential is a rapid, transient decrease in membrane potential (*depolarization*) that can travel for a long distance without loss of signal strength. Action potentials are initiated by the opening of voltage-gated Na channels in axons and muscle fibers. These channels are not found in other cells and are only in the axons of neurons and sarcolemma of myocytes. Thus action potentials are only carried by axons and muscle fibers.

> Depolarization to threshold opens voltage-gated Na channels
> to initiate rapid depolarization in an action potential
> Voltage-gated K channels then open to cause rapid repolarization

For an action potential to fire, a graded potential must change the voltage of the membrane enough to open the voltage-gated Na channels of an axon or muscle fiber. The voltage change must reach *threshold*. In sensory endings, the graded potential might be initiated by a pressure change that opens a mechanically-gated ion channel or by release of a chemical that activates a ligand-gated channel to transmit a pain signal to the spinal cord and brain. Graded potentials in muscle fibers are initiated when an action potential traveling along a motor neuron reaches the neuromuscular junction.

> Action potentials have three phases: resting, depolarization, repolarization
> In resting phase, open K ion leak channels make membrane much more
> permeable to K ions than Na ions; voltage-gated channels are closed

The graded potential spreads out along the membrane to reach the voltage-gated channels of the axon or muscle fiber. If the stimulus is strong enough, the change in membrane potential will exceed the threshold required to open voltage-gated Na channels. Once these channels open, an all-or-none action potential will travel the length of the axon or muscle fiber.

> During depolarization, voltage-gated Na channels open to allow Na ions to enter the cell, causing a reversal of the membrane potential

Opening of *voltage-gated sodium channels* causes Na ions to enter the cell. This switches the membrane potential from negative to positive (*depolarization phase*). Voltage-gated sodium channels are only open for a short time, so the action potential is very brief. To rapidly return the membrane potential to its resting state, *voltage-gated K channels* must open.

> Repolarization begins with inactivation of voltage-gated Na channels and opening of voltage-gated K channels; Na no longer enters cells and K ions exit cell, driven by chemical and electrical gradient

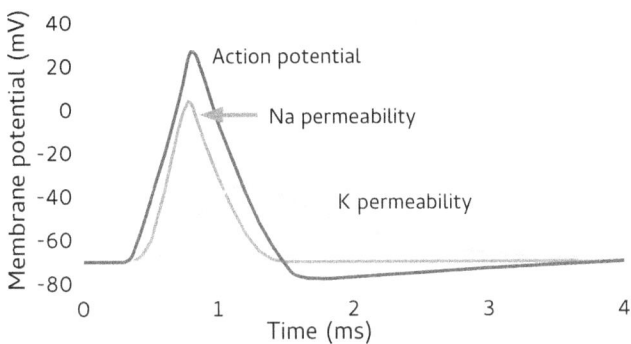

Voltage-gated potassium channels open slowly as the membrane potential becomes less negative during depolarization. This allows K ions to begin leaving the cell. The cell membrane is returned to its resting membrane potential (*repolarization*) by outflow of K ions through these open channels. A slight *hyperpolarization* (more negative membrane potential) is seen at the end of an action potential due to this rapid efflux of K ions from the cell. The voltage-gated potassium channels close after the membrane is restored to its resting membrane potential. Only a very small imbalance of Na and K ions across the membrane results from a series of action potentials.

Many common *antiepileptic drugs* (AEDs) act by blocking voltage-gated Na channels in neurons. These include *phenytoin* and *carbamazepine*. They bind to the inner membrane surface of the inactivated form of the channel, slowing its return to the closed state and prolong-ing the refractory period. This slows the rate of impulse production. Action potentials cannot fire with high frequency, which minimizes the rapid discharge associated with an epileptic episode.

Action Potential Propagation

Explain how an action potential is propagated along an axon or muscle fiber

Action potentials are initiated at the initial segment of an axon, which is close to the cell body in multipolar neurons and the sensory receptor of pseudounipolar neurons. In muscle fibers, the action potential is ini-tiated next to the neuromuscular junction where the motor neuron contacts the muscle fiber.

Voltage-gated Na ion channels open in initial segment causing
depolarization that spreads to open nearby closed channels
Open channels inactivate to prevent backwards propagation

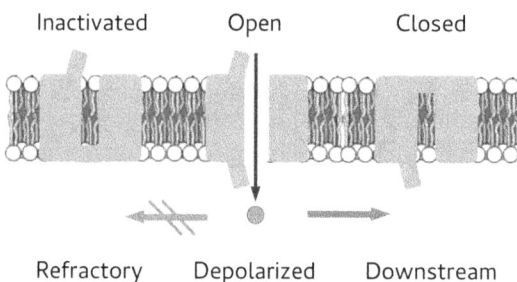

Initiation of the action potential in muscle fibers is the result of open-ing of voltage-gated Na ion channels, which causes depolarization of the membrane. This depolarization passes to downstream closed chan-nels, causing them to reach threshold and rapidly open. Where the action potential has already fired, the voltage-gated Na ion channels are in the inactivated state and cannot open, so propagation is one way, from initial segment along the axon to the axon terminal or from neu-romuscular junction to the each end of the muscle fiber.

Refractory Periods

Explain how refractory periods limit the frequency of firing of action potentials

After opening, voltage-gated Na channels enter the inactivated state. In the *inactivated state*, voltage-gated Na channels cannot be opened, no matter how strong the signal might be. Gradually, inactivated channels return to the closed state that can be opened at threshold voltages. Thus, closed channels can open; open channels inactivate; and, inactivated channels slowly close.

At the peak of the action potential, open voltage-gated Na channels become inactivated and cannot open, even with a strong stimulus. The excitable membrane is in its *absolute refractory period*. During the later *relative refractory period*, most of the voltage-gated Na ion channels are in the closed state, but voltage-gated K ion channels are still open. While K channels are open it is not as easy to depolarize the membrane. A higher threshold stimulus is required to fire another action potential.

> Refractory periods limit frequency of action potentials
> Absolute refractory period: no action potential can fire
> Relative refractory period: stronger stimulus required for firing

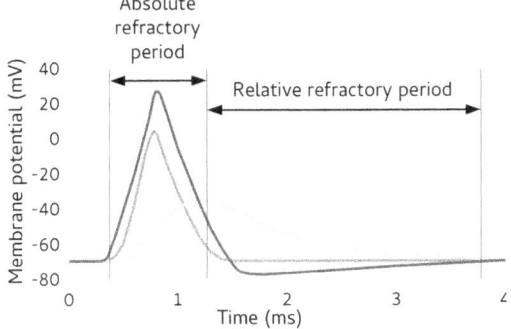

Refractory periods limit the frequency of action potential firing. During the absolute refractory period another action potential cannot fire. During the relative refractory period a stronger stimulus is required to fire another action potential. A waiting period is required before another action potential can fire. Weak signals must wait until the relative refractory period is complete. Strong stimuli can cause more rapid firing (higher frequency action potentials).

Saltatory Conduction

Describe saltatory conduction and the role of myelin

Motor neurons and most sensory neurons are *myelinated*. In the peripheral nervous system myelin is produced by individual *Schwann cells* that wrap around small segments of axons. In the central nervous system (brain and spinal cord), *oligodendrocytes* myelinate nearby neurons. In-between the myelin sheaths are uncovered spaces, the *nodes of Ranvier*.

Voltage-gated channels are found within the nodes of Ranvier and not under the myelin. Thus, myelin is an electrical insulator. Action potentials jump from node-to-node (*saltatory conduction*) and are transmitted much faster along a myelinated neuron than along one that lacks myelin (*unmyelinated*).

> Many neurons are myelinated: myelin sheaths are separated by nodes of Ranvier where voltage-gated channels are located
> Action potentials jump from node-to-node by saltatory conduction

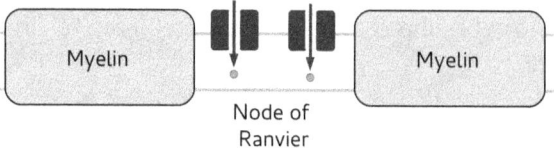

Synapses

Describe how an action potential causes release of neurotransmitter at a synapse

Synapses are sites of communication between axons and postsynaptic cells. Action potentials arriving at a synapse open *voltage-gated Ca channels*, allowing Ca ions to enter the axon terminal. Ca entry into the axon terminal causes release of neurotransmitter from secretory vesicles into the synaptic cleft. Thus, an electrical signal is converted into a chemical signal. The trigger for release of neurotransmitter is Ca ion entry. Neurotransmitters bind to specific receptors to cause a response in the postsynaptic cell

> Synapses link neurons to post-synaptic cells; most use neurotransmitters
> Action potentials arriving at axon terminal open voltage-gated Ca channels
> Ca entry releases neurotransmitter that passes across synaptic cleft

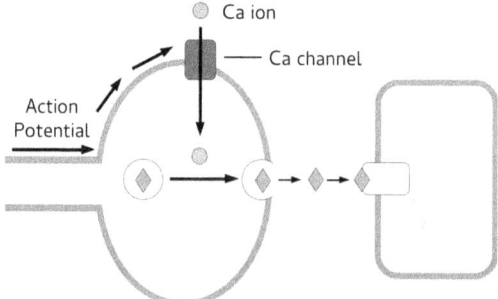

The amount of neurotransmitter released at an axon terminal depends on the frequency of action potentials that reach the synapse. Any individual action potential has exactly the same shape and strength as any other in a given axon. The only way that stimulus strength can be encoded in an axon is by *action potential frequency*. Each time an action potential arrives at a synapse, a small quantity of neurotransmitter is released. Higher action potential frequencies (more rapid trains of impulses) cause greater release of neurotransmitter. Higher neurotransmitter concentrations in the synaptic cleft activate more receptors, causing a stronger response in the post-synaptic cell.

> Strong stimuli increase frequency of action potentials in axons
> Higher action potential frequency increases neurotransmitter release

Motor Units

Describe the components of a motor unit

Nerve impulses travel as electrical signals (action potentials) along *motor neurons* to muscle fibers. Motor neurons are multipolar, myelinated neurons that arise in the central nervous system (brain and spinal cord). A single motor neuron innervates multiple muscle fibers. Together, the motor neuron and all the muscle fibers it innervates is a *motor unit*. Where fine motions are required, such as with the fingers, motor units are small. Each motor neuron innervates a small number of muscle fibers. This gives fine control over movements. Where less fine control is required, such as in walking or standing, motor units are large: many muscle fibers are innervated by each motor neuron.

> Motor unit is one motor neuron and the skeletal muscle fibers it innervates
> Motor neurons arise in CNS and are multipolar and myelinated
> Axon terminals synapse with muscle fibers at neuromuscular junctions

Neuromuscular Junction

Describe the structure and function of the neuromuscular junction

At the muscle fiber, motor neurons form a junction (*neuromuscular junction*) or synapse. When the motor neuron action potential arrives at the neuromuscular junction, it causes release of *acetylcholine* from the axon terminals of the neuron into the synaptic cleft that separates it from the muscle fiber membrane.

> At the neuromuscular junction, acetylcholine is released and passes across
> the synaptic cleft to bind to acetylcholine receptors in the motor end plate
> ACh secretion is triggered by opening of voltage-gated Ca channels

Acetylcholine is produced in the axon terminals of the motor neuron and packaged into secretory vesicles (*synaptic vesicles*) that fuse with the axon terminal membrane to release their contents into the synaptic

cleft. The fusion event is activated by Ca ions that enter the axon terminal through *voltage-gated Ca channels* that open when the motor
neuron action potential activates them. After release, acetylcholine is
degraded to acetate and choline within the synaptic cleft by *acetylcholinesterase* and the choline is recycled to make more
neurotransmitter.

> Binding to ACh receptor directly opens ion channel that allows Na ions
> to enter and depolarize muscle cell at motor end plate
> Acetylcholine is degraded by acetylcholinesterase in synaptic cleft

Acetylcholine binds to specific acetylcholine receptors on the muscle
fiber *motor end plate*. These are nicotinic cholinergic receptors
(ionotropic nACh receptors) just like those in the autonomic ganglia of
the autonomic nervous system. Acetylcholine (ACh) binding opens
cation channels in the sarcolemma, allowing Na ions to enter the muscle fiber and cause local depolarization. This *end-plate potential*
spreads outward to the surface of the muscle fiber where voltage-gated
Na ion channels are located.

ACh receptors are ligand-gated ion channels because binding of the
neurotransmitter directly opens the channel. They are classified as
nicotinic cholinergic receptors because the drug nicotine is an agonist.
Like all nicotinic cholinergic receptors, the response is excitatory. The
end-plate potential is always strong enough to open the voltage-gated
Na channels and initiate a *muscle fiber action potential*. Thus, the electrical signal (action potential) of the motor neuron is transferred to an
electrical signal in the muscle fiber through the neurotransmitter
acetylcholine that passes between the two cells through a synapse.

> Acetylcholine receptor in motor end plate is ionotropic
> Binding directly opens an ion channel that allows Na ion entry
> Na ion entry depolarizes motor end plate (graded potential)

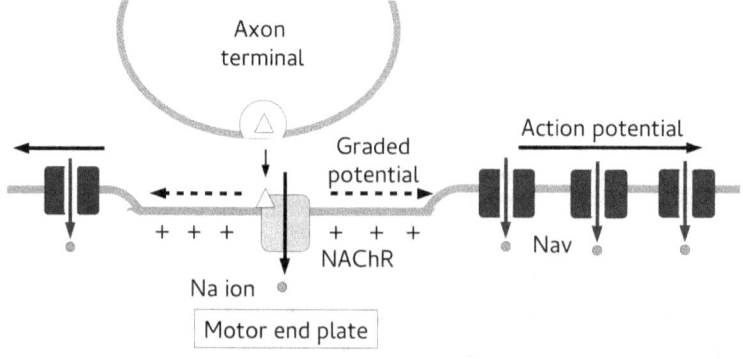

Inhibitors at the Neuromuscular Junction

List agents that alter function at the neuromuscular junction

A number of poisons act at each step of transmission of signals across the neuromuscular junction. Botulinum toxin blocks ACh release from presynaptic terminals. This causes muscular paralysis and death from respiratory failure. *Curare* competes for binding to the nicotinic receptor at the motor end plate which kills is victims by paralysis. A less potent form of the poison (*tubocurarine*) can be used to relax skeletal muscles during anesthesia. *Neostigmine* is an acetylcholinesterase inhibitor that increases ACh concentrations in the synaptic cleft. It is used to treat myasthenia gravis, muscle weakness caused by antibodies that block the ACh receptors.

Review Questions

1. What are the special anatomical features of neurons? How are neurons classified by structure?
2. What is a graded potential and how is it initiated? How is the strength of a graded potential related to Na ion permeability?
3. What are the three states of voltage-gated Na ion channels and how does one transition into another?
4. What are the stages of an action potential and what happens in each stage?
5. How are action potentials propagated along a neuron? What ensures that it travels in one direction?
6. What is the difference between the absolute and relative refractory periods?
7. How do weak and strong stimuli influence action potential frequency?
8. What is myelin? What is a node of Ranvier? What is the mechanism of saltatory conduction?
9. How do action potentials cause neurotransmitter release?
10. What is the relationship between action potential frequency and neurotransmitter release?
11. What is a motor unit? What kind of neurons are motor neurons?
12. What is the neuromuscular junction? What is a motor end plate?
13. What is the effect of acetylcholine binding to its receptor in the motor end plate?

14. How are graded potentials generated in motor end plates of muscle fibers?

15. How are graded potentials converted into muscle action potentials?

Skeletal Muscle Fibers

The last lecture introduced us to the remarkable properties of voltage-gated channels and how voltage-gated Na and K channels can be used to transmit signals long distances along axons and skeletal muscle fibers. We discovered voltage-gated Ca channels in their role in causing release of neurotransmitter from axon terminals in synapses. The lecture finished with the *neuromuscular junction*. Now, the question arises, how does the action potential in a skeletal muscle fiber membrane cause muscle contraction? The key to this is another voltage-gated Ca channel that responds to muscle action potentials. Activation of this *L-type Ca channel* by an action potential causes release of Ca from its stores within the muscle fiber. Once this Ca is released, it triggers an interaction between the thick and thin filaments of the muscle fiber that causes the muscle to contract.

The source of ATP for muscle contraction depends on the intensity and duration of exercise. Quick energy is provided by cleavage of creatine phosphate. Brief, high intensity exercise (sprinting) gets energy from anaerobic metabolism of glucose to lactate. ATP for endurance exercise (walking) is provided by aerobic oxidation of glucose and fatty acids.

Muscle action potentials travel rapidly from the *neuromuscular junction* along the entire length of the muscle fiber membrane (*sarcolemma*). Invaginations of the sarcolemma (*t-tubules*) carry the action potential into the interior of the muscle fiber where they encounter L-type Ca channels. The depolarizing voltage of the muscle action potential activates these voltage-gated Ca channels which are coupled through *ryanodine receptors* to the *sarcoplasmic reticulum* (SR), where Ca is stored in muscle fibers. Activation opens Ca channels in the SR, releasing Ca ions for muscle contraction.

Skeletal Muscles

Describe the general organization of a skeletal muscle

Skeletal muscle cells are referred to as *muscle fibers* because they have an elongated shape. They have multiple nuclei (are multinucleated) and are grouped into bundles (*fascicles*) surrounded by a connective tissue *perimysium. Endomysium* is areolar connective tissue that surrounds individual muscle fibers. The entire muscle is enveloped by *epimysium*.

Fascia is fibrous connective tissue deep to the skin that envelops and separates organs, including muscles. *Deep fascia* refers to connective tissues that envelop muscles. It is continuous with the epimysium that surrounds each muscle and the tendons that connect bones to muscles. Fascia moves along with muscle when the body moves.

> Skeletal muscles consist of muscle fibers (cells) bundled into fascicles
> Muscle fibers are surrounded by endomysium; fascicles by perimysium
> Entire muscle is surrounded by epimysium that blends into tendons

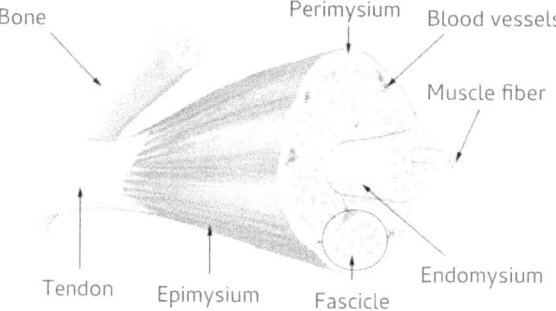

Skeletal Muscle Metabolism

Describe how muscle metabolism changes with activity and exercise

Resting muscle and *low-intensity exercise* use aerobic metabolism to produce ATP. In the presence of oxygen, nutrient sources for muscles include glucose, stored glycogen that is converted to glucose, and fatty acids. Glucose and glycerol from breakdown of triglycerides are metabolized to pyruvate before entering the mitochondria to produce abundant ATP by oxidative phosphorylation. This ATP is used for metabolic activities in the cells and to restore glycogen and creatine phosphate stores.

> Resting muscle produces creatine phosphate and stores glycogen
> High-intensity bursts of effort use creatine phosphate to produce ATP
> Active muscle uses blood glucose and mobilizes glucose from glycogen

Exercising muscle first recruits *creatine phosphate*, which exchanges its phosphate group with ADP to provide a short burst of ATP for high-intensity bursts of effort, such as lifting a weight. ATP for longer periods of high-intensity exercise (running) is provided by glycolysis (*anaerobic metabolism*), using a combination of blood glucose and stored glycogen.

> High-intensity exercise uses anaerobic metabolism to produce lactate
> Endurance exercise uses blood glucose and fatty acids
> Aerobic metabolism produces large amounts of ATP

Endurance exercise is called aerobic because it uses aerobic metabolism of blood glucose and fatty acids, much like resting muscle. Thus, muscle metabolism shifts according to the intensity and duration of exercise.

> Short bursts of high-intensity activity get energy from creatine phosphate
> Then glycogen is hydrolyzed to glucose for anaerobic glycolysis
> Endurance exercise utilizes aerobic metabolism of glucose and fatty acids

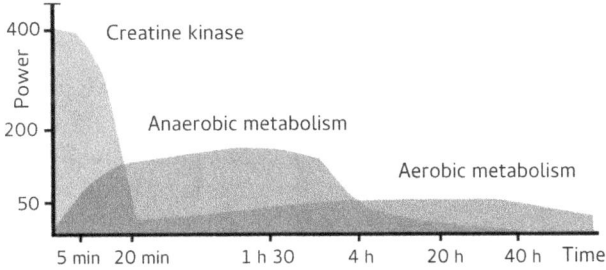

Muscle Fiber Types

Distinguish muscle fiber types and their metabolism

Skeletal muscle twitches may be fast or slow. Muscle fibers are divided into slow twitch and fast twitch classes. Contraction of cardiac and smooth muscle cells are slower than even the slowest skeletal muscle contraction. *Slow twitch fibers* are used in postural muscles that remain contracted for long periods of time. They use aerobic metabolism. Fast twitch fibers are used for rapid movements, such as blinking the eyelids. Most fast twitch fibers use anaerobic metabolism.

The difference between the two fiber types is evident by light microscopy. Muscle fibers contain the red pigment myoglobin that binds oxygen, which can be used to support aerobic metabolism. *Slow oxidative* (type I) fibers have abundant myoglobin and mitochondria and use aerobic metabolism for endurance. *Fast glycolytic* (type II) fibers have few mitochondria, use anaerobic metabolism during contraction, and are easily fatigued. Both fiber types are found together in skeletal muscles, but glycolytic fibers dominate in fast-twitch muscles and aerobic fibers in slow-twitch muscle.

> Slow oxidative (type I) muscle fibers have red myoglobin pigment
> that binds oxygen to support aerobic metabolism
> Fast glycolytic (type II) muscle fibers have abundant glycogen

Fatigue-resistant, slow-twitch muscle fibers (Type I) are recruited first in a movement, producing minimal force. Higher levels of stimulation recruit fast-twitch, oxidative and glycolytic fibers (Type IIA and IIB) which generate more force. Near maximum force is achieved by recruiting glycolytic fast-twitch muscle fibers that fatigue rapidly. Fatigue is minimized by asynchronous recruitment of different motor units in the same muscle to maintain muscle tone: one motor unit fires and rests, then another, and another, cycling back to the first motor unit which has now recovered.

Muscle Twitch

Define the stages and events of a muscle twitch

A *muscle twitch* is the response of a skeletal muscle to a single stimulation. During the *latent period*, the signal passes from the motor neuron through the neuromuscular junction and along the sarcolemma into the muscle fiber through t-tubules to cause Ca release. During the *contraction period*, tension increases as the sarcomeres shorten. In the *relaxation period*, tension is reduced as the muscle fibers return to resting length. The strength of a contraction is measured by the tension (pull) produced by the muscle when it is attached to a fixed measuring device.

> Muscle twitch is response of muscle to a single stimulation
> Latent period: transfer of action potential from motor neuron to Ca release
> Contraction period: shortening of muscle; relaxation: return to rest

Repeated contractions increase muscle tension (*wave summation*). This is due to increased numbers of muscle fibers participating in the contraction. To increase the number of muscle fibers that fiber, additional motor neurons must fire (*motor unit recruitment*). *Tetanus* is continuous muscle contraction due to rapidly repeated stimuli. It is the maximum contraction that can be obtained by a muscle.

> Repeated stimulation of skeletal muscles increases tension produced
> Rapidly repeated stimuli fuse to produce maximum contraction (tetanus)

Fatigue is relaxation of a muscle even though it continues to be stimulated. It is due to several factors, including depletion of energy reserves and accumulation of metabolites. *Muscle tone* refers to ongoing cycling of muscle fiber contractions from one motor unit to another. This cycling minimizes fatigue. Fatigue is also lessened if only a small number of motor units are recruited in a movement. Thus, repetitive movements should be done in as relaxed a fashion as possible. Fatigue is more rapid if a muscle is contracted to a maximal extent. This is why lifting a heavy weight causes you to more rapidly tire than would lifting a light weight.

Muscle Fibers

Explain the structure of a muscle fiber and its t-tubules

Muscle fibers are skeletal muscle cells. The components of a muscle fiber are given special names. The cytoplasm of a muscle fiber is its *sarcoplasm*. The plasma membrane is its *sarcolemma*. T-tubules (*transverse tubules*) are extensions of the sarcolemma that penetrate deeply into the muscle fiber. The *sarcoplasmic reticulum* (SR) is a network of membranes, equivalent to the endoplasmic reticulum of other cell types, that surrounds each myofibril. *Cisternae* are expanded regions of the SR that store calcium. Together, a T-tubule and its adjacent pair of cisternae form a *triad*.

Muscle fibers are multinucleated cells; their sarcolemma (cell membrane) penetrates into fibers as t-tubules that contact terminal cisterna of Ca-containing sarcoplasmic reticulum (SR) at triads

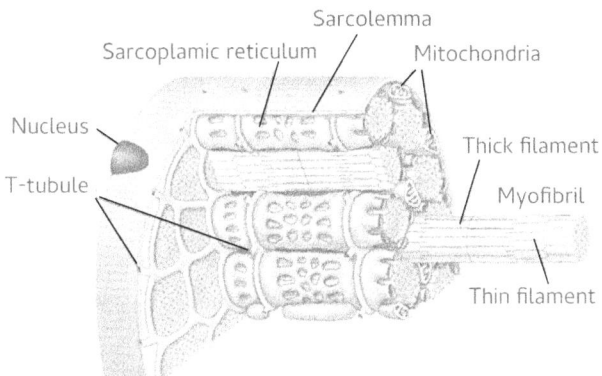

When a motor neuron fires an action potential, it travels down the neuron to the muscle fibers it innervates. There, acetylcholine is released to initiate an action potential in the muscle fiber. The muscle fiber action potential is carried along the muscle fiber membrane and into the muscle fiber through t-tubules. At the triads, the action potential of the t-tubules causes the release of Ca ions. If ATP is available, the Ca ions initiate muscle contraction.

Excitation-Contraction Coupling

Describe the events from motor neuron to calcium release

Muscle contraction requires ATP and calcium ions. ATP supplies the energy and Ca ions supply the trigger to initiate contraction. Ca is released in response to action potentials sent to muscle fibers by motor neurons. Motor neuron action potentials open voltage-gated Ca ion channels in axon terminals which cause exocytosis of acetylcholine into the synaptic cleft. Binding of acetylcholine to specific post-synaptic receptors in the motor end plate open cation channels that allow entry of Na ions into the muscle fiber.

> Depolarization at motor end plate (end plate potential) spreads to voltage-gated Na channels of sarcolemma to initiate muscle action potential which enters muscle fibers through t-tubules to release Ca from SR

The local potential (*end-plate potential*) initiated by opening of nicotinic cholinergic receptor cation channels in the motor end plate spreads to the sarcolemma (muscle fiber membrane) where voltage-gated Na and K channels are located. The properties of the sarcolemma are much like those of an axon. It can propagate action potentials along the membrane. As in an axon, local depolarization opens voltage-gated Na channels to initiate a *muscle action potential* that travels all along the muscle fiber sarcolemma without loss of signal strength.

> Muscle action potentials arriving at triads activate L-type Ca channels
> L-type Ca channels open ryanodine receptors that release Ca from SR
> Ca is constantly taken back into SR by Ca ATPase

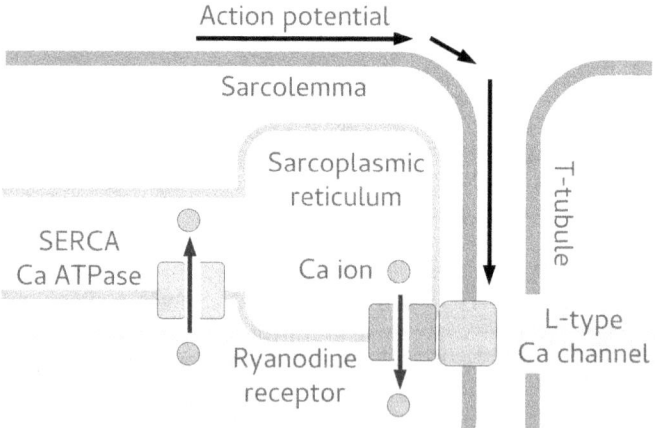

How does the action potential enter the muscle fiber to cause release of calcium from the sarcoplasmic reticulum? This problem is *excitation-*

contraction coupling. The action potential enters the muscle fiber, as an electrical signal, through t-tubules that are extensions of the sarcolemma. When the muscle action potential reaches the triad, depolarization activates *L-type Ca channels*. The "L" stands for long-lasting: the channel stays open for a relatively long time. They are often called DHP channels because they are inhibited by drugs of the dihydropyridine class. L-type Ca channels in skeletal muscle fibers are directly coupled to *ryanodine receptors* in the sarcoplasmic reticulum membrane. Activation of L-type channels in the t-tubules open ryanodine receptors which cause release of Ca ions from the sarcoplasmic reticulum. This release of Ca initiates muscle contraction. Ca is constantly returned to the sarcoplasmic reticulum by the ATPase SERCA (sarco/endoplasmic reticulum Ca-ATPase).

Sarcomeres

Describe the organization of a sarcomere and its myofibrils

Muscle fibers (cells) contain myofibrils (thick and thin filaments) organized as contractile *sarcomeres*. *Myofibrils* are groups of contractile proteins arranged as long fibrils inside muscle cells. They are composed of multiple myofilaments. *Myofilaments* are overlapping thick and thin filaments within each myofibril. *Thin filaments* are attached to Z lines that define the ends of sarcomeres. Dark striations within sarcomeres are A bands (A = anisotropic) and contain *thick filaments*. Light strips between striations within sarcomeres are I bands (I = isotropic) and contain only thin filaments.

Myofibrils are made up of overlapping thick and thin filaments
(myofilaments) that form dark (A) and light (I) bands
Thin filaments attach to Z lines at the ends of contractile sarcomeres

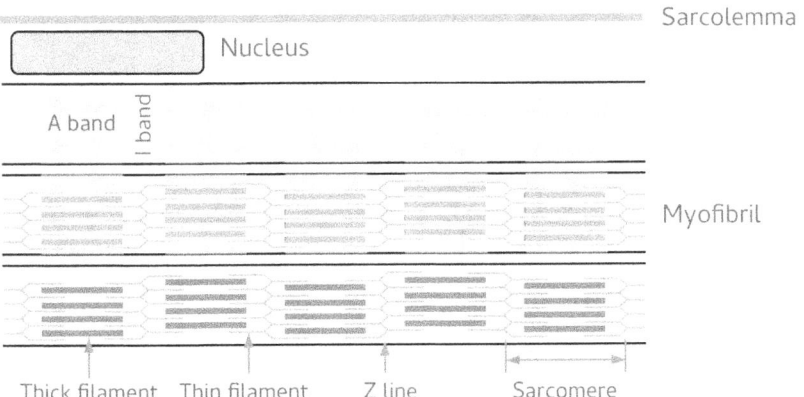

Sliding Filament Model of Muscle Contraction

Describe the sliding filament model of muscle contraction

In the center of the sarcomere is an H zone that has only thick fila-ments. Muscle contraction shortens the sarcomere, decreasing the distance between Z lines. The A band remains constant because the thick filaments do not move. The I band and H zone get smaller during contraction. The pulling together of Z lines by overlapping thick and thin filaments is the *sliding filament model* of muscle contraction.

> Sliding filament model: thin filaments slide over thick filaments
> Sarcomere shortens: A band does not change and I band shortens

Relaxed

Contracted

Thin and Thick Filaments

Describe the structure of thick and thin filaments

Thick filaments consist of myosin protein molecules connected to-gether by association of their long tails. *Myosin heads* decorate the surface of the thick filaments. *Thin filaments* have a more complex structure. The basic thin filament is composed of globular G-actin units associated together into a double helix of F-actin (filamentous actin). Covering the helix are *tropomyosin* molecules that block myosin bind-ing sites on the G-actin subunits. *Troponin* is a globular, Ca binding protein associated with thin filaments.

> Myosin heads attached to tails associate to form thick filaments
> G-actin associates into double helix covered by tropomyosin
> Ca binding to troponin shifts tropomyosin to expose myosin binding sites

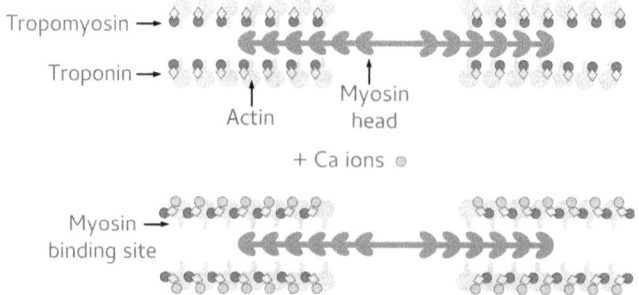

Tropomyosin →

Troponin →

Actin

Myosin head

+ Ca ions ⊙

Myosin → binding site

Ca binding to troponin shifts the position of tropomyosin to expose myosin binding sites on G-actin. So long as Ca is available (when the motor neuron is firing), myosin binding sites are exposed. When action potentials end, intracellular Ca levels decrease and tropomyosin returns to its resting position. Blocking of myosin binding sites by tropomyosin prevents cross-bridge formation between myosin heads on thick filaments and G-actin of thin filaments. Ca binding to troponin allows cross-bridge formation to initiate muscle contraction.

Cross-bridge Cycle

Describe the cross-bridge cycle of skeletal muscle contraction

ATP hydrolysis provides energy to rotate myosin heads into a cocked position before they bind to actin. When calcium is available, myosin can bind to actin where thick and thin filaments overlap. During muscle contraction, myosin heads on the thick filaments bind to actin on the thin filaments and pull them toward the center of the sarcomere. In *cross-bridge formation*, the cocked myosin heads bind to the exposed sites on actin. As the cocked myosin heads rotate, they pull the thin filaments toward each other in the *power stroke.*

ADP is released at the end of the power stroke, as myosin heads unbinds from thin filaments. ATP then binds to myosin heads, causing them to cock once again in preparation for repeating the cycle of cross-bridge formation and rotation. The cycle of cross-bridge formation, contraction, and ATP cleavage continues until nerve impulses cease.

> ATP cleavage cocks myosin head; Ca exposes myosin binding sites
> Cocked myosin head binds to actin when (cross-bridge formation)
> Rotation of myosin head pulls thin filaments toward each other (power stroke)

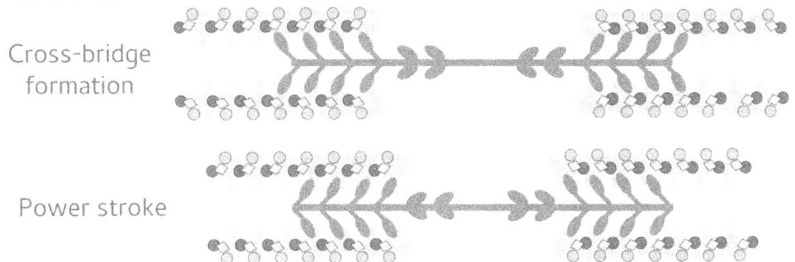

Cross-bridge formation

Power stroke

Calcium and ATP are both required for contraction. ATP is supplied by aerobic metabolism when oxygen is abundant and by anaerobic metabolism when it is limited. Calcium is the trigger for contraction and is released from the sarcoplasmic reticulum so long as motor nerve impulses are active. When nerve impulses stop, calcium is no longer released from the cisternae, so calcium levels drop and tropomyosin again covers the myosin binding sites on actin causing relaxation of the muscle fiber.

Length-Tensian Relationship

Explain the relationship between initial sarcomere length and strength of muscle contraction

In skeletal muscles, increased tension can be developed by recruiting more motor units, causing more muscle fibers to contract at the same time. In addition, the tension developed depends on the initial length of the muscle. When myofilaments optimally overlap, the greatest possible tension can be produced. When the muscle is stretched, thin and thick filaments do not overlap optimally and fewer cross-bridges can form when the muscle begins to contract and the tension produced is less. Similarly, tension development is less when the muscle is overly short. Thus, when lifting a weight with the forearm, maximum force is generated when the forearm is near its middle position. If a muscle is overstretched in the lab, passive tension due to connective tissue resistance dramatically increases.

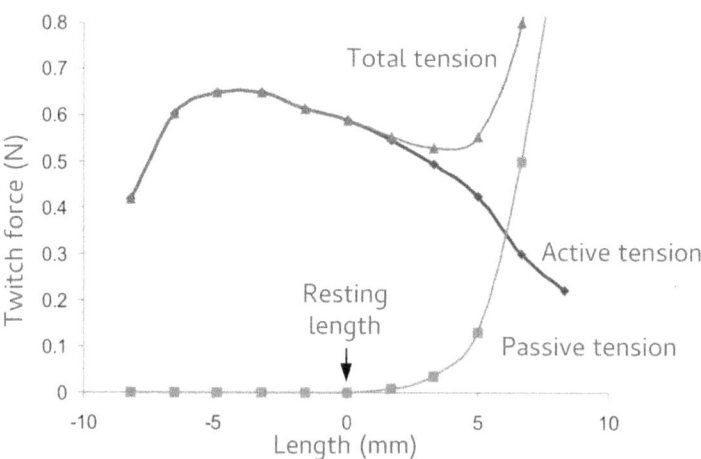

Maximum tension produced with optimal overlap of thin and thick filaments
Passive component of tension dramatically increases when muscle stretches

Review Questions

1. Distinguish endomysium, perimysium, and epimysium. What is a muscle fascicle?

2. What is meant by a muscle fiber?

3. What is the role of creatine phosphate in muscle contraction?

4. How do the sources of ATP change as intensity of muscle contraction and duration of exercise change?

5. What are the two main skeletal muscle fiber types and how do they differ?

6. What are the stages of a muscle twitch?

7. What happens when a skeletal muscle is stimulated over and over?

8. What is the relationship of sarcolemma to t-tubule? What is a triad?

9. List the steps of excitation-contraction coupling in skeletal muscles.

10. How do action potentials enter muscle fibers? How do they cause Ca release?

11. Describe the internal structure of a muscle fiber.

12. How do overlapping filaments in sarcomeres explain the striped appearance of skeletal muscle?

13. What is a sarcomere? To what do thin filaments attach?

14. Describe the sliding filament model of muscle contraction.

15. What is the relationship between myosin molecules and thick filaments?

16. What are the components of a thin filament and how are they arranged?

17. What happens when Ca binds to thin filaments?

18. What is the cross-bridge cycle? What drives the power stroke?

19. What is the role of ATP in muscle contraction?

20. How does muscle tension change as a muscle is stretched? What explains the relationship between length and tension?

Cardiovascular Physiology

The animal's heart is the basis of its life, its chief member, the sun of its microcosm; on the heart all its activity depends, from the heart all its liveliness and strength arise.

– William Harvey

Cardiac Physiology

The study of receptors and chemical messengers, membrane transport, and the properties of excitable tissues form the basic foundation for understanding the physiology of the body. We have seen how specific molecules carry out their functions by binding to receptors that alter cell function, how transport of materials across cell membranes is controlled by specific transporter and channel molecules, how signals are carried long distances to specific locations by neurons, and how muscles contract. Throughout our study we observed that body functions are compartmentalized. They occur in specific cells or compartments (such as intracellular or extracellular, in mitochondria or the nucleus).

With this strong foundation, we can now turn to study organ systems and how they function. We begin with the cardiovascular system because it is the one most likely to require attention in the clinic. Nearly half of all adults have hypertension according to the new criteria of 130/80 as target maximum blood pressure. The truism that a healthy mind makes a healthy body (me*ns sana in corpore sano*, Juvenal*)* also works in reverse. A strong cardiovascular system maintained by adequate exercise keeps the brain healthy, as well. The second line of Juvenal's poem is "ask for a stout heart."

The heart is a rather simple organ, consisting of four *chambers* linked by *valves* that receive blood and pump it. This first lecture on the heart will focus on this basic function, examining what causes the valves to open and close, how the ventricles are efficiently filled, and the steps that occur during a single heartbeat (the *cardiac cycle*). We get into the weeds in the following lecture, which discusses how the heart changes its rate of beating and strength of contraction, topics that will require all our knowledge of hormones, autonomic nervous system, channels, transporters, ion fluxes, and muscle contraction.

Cardiovascular System

Describe the overall organization of the cardiovascular system

The *cardiovascular system* includes heart, blood, and vessels which form a continuous, closed circulation. The *heart* is a dual pump that circulates blood in a continuous circuit through systemic and pulmonary tissues. The *systemic circulation* is from left ventricle (LV) through the body to the right atrium (RA). The *pulmonary circulation* is from right ventricle (RV) through the exchange surfaces of the lungs to the left atrium (LA).

Cardiovascular system: heart, blood, and vessels
Systemic circulation: left ventricle through body to right atrium
Pulmonary circulation: right ventricle through lungs to left atrium

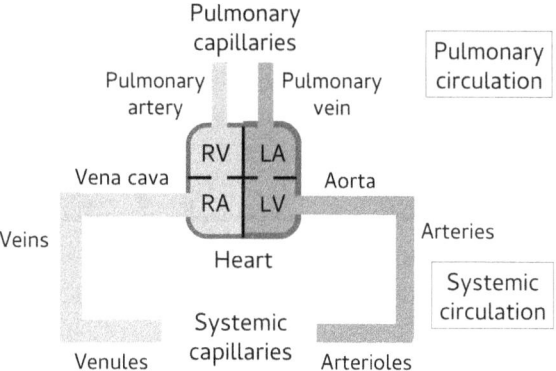

An *artery* carries blood away from the heart; small arteries are *arterioles*. The aorta is an artery that carries oxygenated blood to the body and pulmonary arteries carry deoxygenated blood from heart to lungs. A *vein* carries blood toward the heart. Small veins are *venules*. The superior and inferior vena cavae return deoxygenated blood from the body to the heart. Pulmonary veins carry oxygenated blood from lungs to heart. Between arterioles and venules are beds of *capillaries*: thin-walled blood vessels that allow passage of extracellular fluids between blood and interstitial fluids that bathe the cells.

The cardiovascular system provides nutrients to and removes wastes from tissues. Functions of blood include transport, regulation, defense, and homeostasis. Blood transports gases, nutrients, hormones, and wastes. Blood regulates the pH and ionic composition of interstitial fluids. Blood cells and proteins transported in the blood defend against toxins and pathogens. Blood coagulation minimizes fluid loss in injury. Blood helps stabilizes body temperature by redistribution of blood flow.

Anatomy of the Heart

Describe the gross structural features of the heart

The *apex* of the heart is its inferior, pointed end that rests on the diaphragm. The base of the heart is its superior, flat end from which the major blood vessels arise. The *mediastinum* is the central compartment of the thoracic cavity that contains the heart and its blood vessels. The *coronary sulcus* is a surface depression between the atria and ventricles. The anterior and posterior interventricular sulci are between the ventricles on the surface of the heart.

Heart is centrally located within the mediastinum
Great vessels emerge from its base (superior)
Tip of left ventricle is apex (inferior)

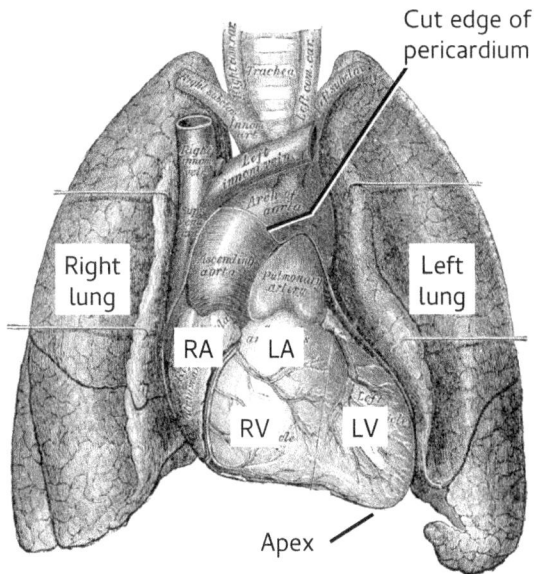

Fibrous pericardium is the outermost layer of dense irregular fibrous tissue covering the entire heart. Beneath it is the serous *pericardium*, a double-layered membrane directly surrounding the heart. The outer layer of this double membrane is *parietal pericardium*, which is fused to the fibrous pericardium. The *epicardium* is the inner layer of the serous pericardium, so may also be called visceral pericardium. *Pericardial fluid* fills the space between the two layers of the pericardium in the *pericardial cavity*, allowing the heart to slip within its sac while it beats.

Heart is enclosed in pericardium (serous double membrane)
Pleural cavity between parietal and visceral pericardium
contains pleural fluid that allows heart to slip as it beats

The heart wall proper consists of three layers. Epicardium forms the outermost layer. Directly beneath it is the *myocardium*. This middle layer of cardiac muscle tissue makes up the bulk of the heart. The *en-*

docardium is a thin, inner lining layer of endothelium and connective tissue that covers the interior of the heart chambers and its valves. This smooth lining prevents coagulation of blood.

Epicardium: outermost layer of heart wall (visceral pericardium)
Myocardium: thick, middle layer of cardiac muscle tissue
Endocardium: inner lining layer of endothelium and connective tissue

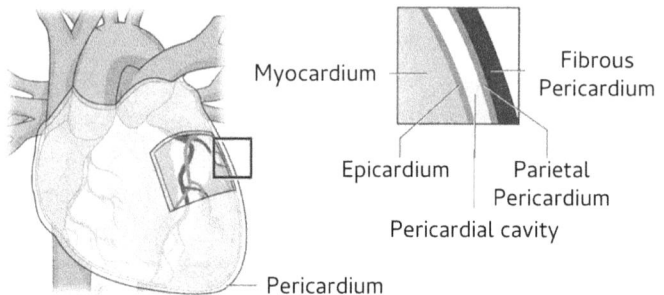

Pericarditis is inflammation of the pericardium which is often associated with sharp chest pain. *Endocarditis* is an inflammation of the endothelium that can damage and destroy the valves of the heart and may spread to the myocardium. It is most often the result of an infection. Endocarditis causes lesions (vegetations) where microbes, platelets, and fibrin adhere that disrupt blood flow and valve function.

Chambers of the Heart

Describe the flow of blood through the heart

The upper chambers of the heart that receive blood are *atria*. The lower chambers that pump blood are *ventricles*. The wall of the left ventricle is thicker than that of the right ventricle, as it must pump against the higher pressure of the systemic circulation. Thus, the *right heart* pumps deoxygenated blood to the lungs and its ventricle has thin walls. The *left heart* pumps oxygenated blood to the system and has thick walls.

Atria: upper chambers receive blood; ventricles: lower chambers pump blood
Left ventricle pumps to body; right ventricle pumps to lungs
Left ventricle wall thicker: pumps against higher systemic pressure

Blood flow through the heart begins with the right heart which receives blood from the body and sends it to the lungs. At the same time, the left heart receives blood from the lungs and pumps it into the body. In the right heart, systemic blood enters the the *right atrium*. It passes through the *tricuspid valve* into the right ventricle. Blood is pumped by the right ventricle through the *pulmonary valve* and into the pulmonary trunk which splits into the left and right pulmonary arteries which supply the air exchange surfaces of the lungs with deoxygenated blood. In the left heart, pulmonary blood enters the *left atrium*. It passes through the *mitral (bicuspid) valve* into the left ventricle. Blood is pumped by the left ventricle through the *aortic valve* into the *aorta* to supply the body with oxygenated blood.

> Blood enters atria, passes through AV valves and enters ventricles
> Tricuspid valve: right AV valve; mitral valve: left AV valve
> Ventricles pump blood through pulmonary valve and aortic valve

Heart Valves

Describe the structure and differences between the heart valves

Between the atria and ventricles are *atrioventricular (AV) valves*. The *tricuspid valve* separates the right atrium from the right ventricle. The *mitral (bicuspid) valve* separates the left atrium from the left ventricle. These valves are set in fibrous rings that form the *cardiac skeleton* to which the myocardium inserts. *Semilunar valves* are between the ventricles and the major arteries leaving the heart. The valve between the right ventricle and pulmonary trunk is the *pulmonary (semilunar) valve*. The valve between the left ventricle and aorta is the *aortic (semilunar) valve*.

Stenosis is a narrowing of the heart valve opening that is often associated with its thickening along the valve edges. Stenosis of the mitral valve slows filling of the left ventricle during diastole (when the myocardium is relaxed and the heart is filling). Aortic stenosis increases the pressure required to eject blood from the left ventricle during systole (when the heart is contracting). Aortic stenosis is the most common disease of the heart valves.

> Atrioventricular valves separate atria from ventricles
> Semilunar valves regulate openings to aorta and pulmonary trunk
> Tricuspid valve: right heart; mitral (bicuspid) valve: left heart

One-way valves ensure that blood flow is in one direction. In atrioventricular valves, strong, tendon-like cords (*chordae tendineae*) are attached to the leaflets (cusps) of the valves. The chordae tendineae prevent the AV valves from inverting when the ventricles contract. *Papillary muscles* attached to the chordae tendineae contract with the ventricles to prevent back-flow into the atria. Semilunar valves have cusps with a half-moon shape. After the ventricles push blood through these valves into the pulmonary trunk and aorta, they snap shut as the blood flows backward during ventricular relaxation.

> Papillary muscles attached to chordae tendineae contract with ventricles to keep AV valves from inverting and prevent back-flow into atria
> Semilunar valves have cusps that snap shut during ventricular relaxation

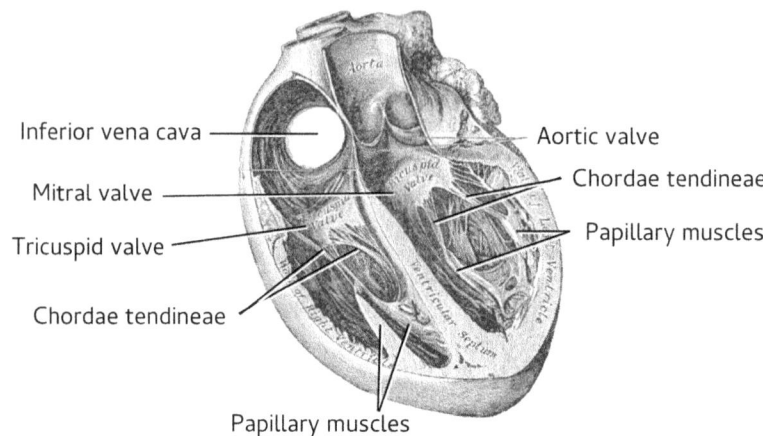

Inferior vena cava — Aortic valve

Mitral valve — Chordae tendineae

Tricuspid valve — Papillary muscles

Chordae tendineae

Papillary muscles

Mitral valve prolapse is often the result of weakened leaflets that may lead to rupture of the chordae tendineae. The valves do not close properly and the valve leaflets enter the atrium during ventricular contraction (systole). This causes regurgitation of blood from the ventricle back onto the atrium.

Cardiac Cycle

Describe the phases of the cardiac cycle

The *cardiac cycle* is the cycle of contraction and relaxation of the atria and ventricles during a complete heartbeat. *Diastole* is the time when the heart is relaxed and the chambers are filling with blood. *Systole* is the time when the heart is contracting and pumping blood. Contraction causes a rise in pressure in blood vessels that is felt as a *pulse* in surface arteries. The heart beats about 70 times per minute.

The cardiac cycle begins with passive *ventricular filling*. AV valves are open and semilunar valves are closed. Blood enters the atria and ventricles due to venous pressure. Then the atria contract. This completes the filling of the ventricles (diastole). The volume of blood in each ventricle after complete filling (the maximum volume) is the *end-diastolic volume* (EDV). Once the ventricles are filled, they begin to contract. Ventricular contraction pushes against the leaflets of the AV valves, which close. This closing of the AV valves causes some turbulence which is registered as the *first heart sound* (S1, lub). Thus S1 marks the start of systole.

> Cardiac cycle: contraction and relaxation during one heartbeat
> Diastole: chambers fill with blood
> Systole: ventricles contract to eject blood

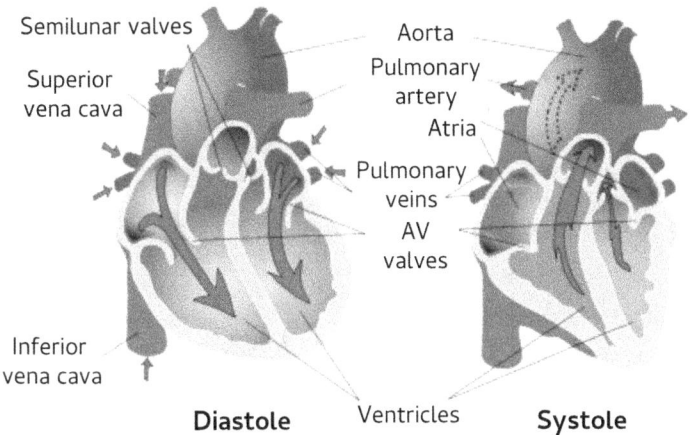

Semilunar valves   Aorta
                        Pulmonary
Superior   artery
vena cava   Atria
                        Pulmonary
                        veins
                        AV
                        valves
Inferior
vena cava

Diastole   Ventricles   **Systole**

At first, pressure in the ventricles is not high enough to open the semilunar valves. Thus, ventricular volume does not change (the valves are closed). It is an *isovolumetric ventricular contraction*. Once ventricular contraction builds enough pressure in the ventricles, the semilunar valves open. When the aortic and pulmonary (semilunar) valves open, blood is ejected from the ventricles into the great arteries (*ventricular ejection*).

When ventricular pressure falls due to this ejection of blood, semilunar valves close, causing the *second heart sound* (S2, dub), which marks the beginning of diastole. As ventricular pressure drops, the heart chambers fill with blood and another cardiac cycle begins. The volume after contraction is the *end-systolic volume* (ESV). This is the volume left in the ventricles after ejection. ESV is the minimum volume of blood in the ventricles during the cardiac cycle.

> EDV: end-diastolic volume; ESV: end-systolic volume
> Stroke volume (SV = EDV – ESV): volume pumped with one heartbeat
> Cardiac output (CO = SV x HR): volume pumped per minute

Stroke volume (SV) is the volume of blood ejected per heartbeat: SV = EDV – ESV. *Heart rate* (HR) is the number of times the heart beats in one minute. *Cardiac output* (CO) is the volume of blood ejected by the ventricles per minute, so it is the product of stroke volume and heart rate (CO = SV x HR). The efficiency of pumping is measured by the *ejection fraction*, which is the stroke volume divided by the maximum filling volume (EDV) expressed as a percentage (SV/EDV x 100%). Typical values are: EDV = 130 mL; ESV = 60 mL; HR = 70 bpm. Thus, SV = 130 − 60 = 70 mL; CO = 70 x 70 = 4.9 L/min; and, EF = 70 / 130 = 54%.

> Ventricular filling: AV valves open, ventricles passively fill, then atria contract
> Isovolumetric contraction: AV valves close, pressure builds
> Ventricular ejection: pressure opens semilunar valves and blood is ejected

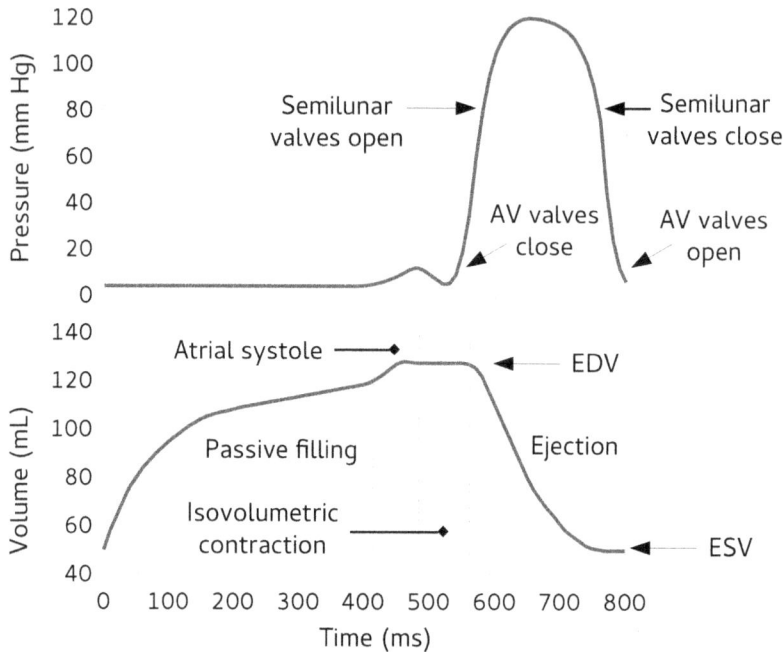

Cardiac Conduction System

Describe the cardiac conduction system and normal sinus rhythm

The cardiac cycle begins with passive filling of the ventricles driven by the pressure of venous return. After the ventricles passively fill with blood, contractions of the atria arise from the *SA (sinoatrial) node* located in the right atrium where the superior vena cava enters. The SA node spontaneously fires action potentials. These activate adjacent cardiomyocytes to fire action potentials in turn. Contractions pass downward and through conducting fibers of the interatrial band (Bachmann's bundle) to the left atrium, squeezing additional blood into the ventricles.

Depolarization spreads over the atria until it reaches the nonconductive fibrous ring at the level of the heart valves. There the signal is channeled into the *atrioventricular (AV) node,* which lies within the interventricular septum just inferior to the fibrous ring. In this way, the AV node pacemaker is driven at a rate set by the SA node pacemaker. In *normal sinus rhythm,* the AV node fires about 0.1 second after atrial depolarization reaches it. This AV nodal delay ensures complete atrial contraction prior to ventricular depolarization.

Signals from the AV node are conducted through the *AV bundle (of His)* and its right and left branches within the interventricular septum. *Purkinje fibers* in the walls of the ventricles conduct the signal upward from the apex of the heart. The spread of the electrical signal causes the ventricles to squeeze blood up and out the arteries.

> Sinoatrial (S) node: pacemaker (initiates heartbeat)
> Depolarization spreads over atria to atrioventricular (AV) node
> Passes through bundles of His and up Purkinje fibers in ventricles

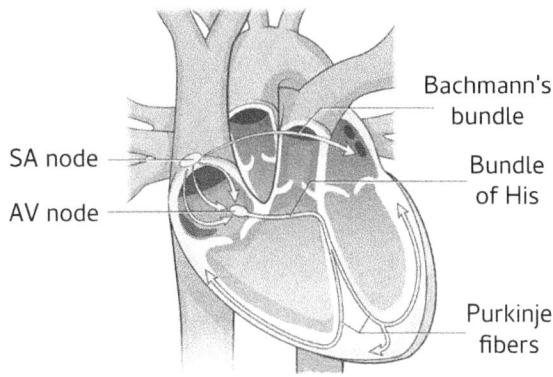

The conducting fibers of the cardiac conduction system are large, modified cardiac muscle cells with few myofibrils that do not contract. Impulse propagation is rapid and is isolated from surrounding tissues by collagen that blocks the spread of impulses away from the conduction fibers.

Electrocardiogram (ECG)

Describe the waves of the ECG

The *electrocardiogram* (ECG) measures the electrical activity of the heart. The ECG has three characteristic waves: the P wave, QRS complex, and T wave. The *P wave* of the ECG is due to contraction (depolarization) of the atria (atrial systole). The *QRS complex* of the ECG marks ventricular depolarization (systole). Atrial depolarization is buried under the QRS complex. The *T wave* of the ECG is due to ventricular relaxation (diastole). The horizontal zero line between waves, which is best seen from the end of P to start of T wave, is the *isoelectric line*.

An interval is the time between two peaks. A segment is the normally isoelectric line between the end of one peak and the beginning of another. Intervals may be shortened or prolonged in heart disease and segments may be elevated or depressed.

The *RR interval* (time between R peaks) is used to measure heart rate. When the RR interval is measured in seconds HR = 60 / RR. The *PR interval* is the time from the beginning of the P wave to the onset of the QRS complex. It represents the time required for the SA pacemaker signal to reach the apex of the ventricles. Normal values are 0.12 to 0.20 seconds. It is prolonged in first-degree AV block.

The *QT interval* is from the onset of QRS to the end of the T wave. QT intervals must be corrected for heart rate. It represents the time for depolarization and repolarization of the ventricles. Corrected QT interval (*QTc*) is calculated by dividing the QT interval by the square root of the RR interval. Long QT syndrome is the result of delayed repolarization and is a risk factor for sudden death.

P wave: atrial depolarization; P-R interval: time from SA node to ventricles
QRS complex: ventricular depolarization; R-R interval represents heart rate
T wave: ventricular repolarization; Q-T interval: ventricular contraction time

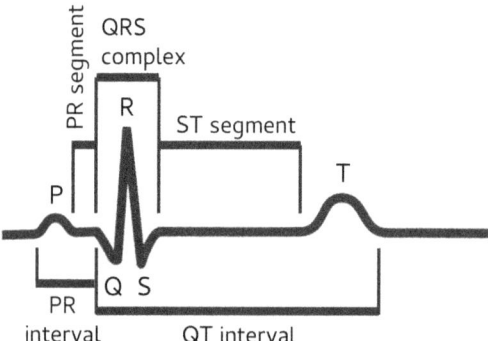

Any disturbance of normal sinus rhythm is an *arrhythmia*. Arrhythmias often arise from ectopic beats, extra beats originating outside SA or AV nodes. *Bradycardia* is slow heart rate (below 60 bpm) and causes dizziness, nausea, vomiting, hypotension, and syncope due to decreased cardiac output. *Tachycardias* are elevated heart rates (above 100 bpm). Those that originate above the ventricles are *supraventricular tachycardias* and include atrial fibrillation, atrial flutter, and paroxysmal tachycardia. *Fibrillation* is uncoordinated contraction. The AV node serves as a "gatekeeper" between atria and ventricles: it stops atrial fibrillation from being transferred to the ventricles.

In *atrial flutter,* the atria contract at a faster rate (often 2:1) than the ventricles due to a blockage of conduction through the AV node. Paroxysmal means from time to time, so *paroxysmal atrial flutter* occurs occasionally, but may last for hours or days. Supraventricular tachycardias reduce ventricular filling and may cause palpitations, lightheartedness, sweating, shortness of breath, and chest pain. Atrial fibrillation is the most common serious abnormal heart rhythm. It is largely controlled by agents that reduce heart rate, such as beta-blockers (metoprolol) and channel blockers (diltiazem).

Ventricular tachycardias are often life-threatening. In *ventricular fibrillation,* the ventricles quiver and cease contracting, which severely compromises cardiac output that often leads to cardiac arrest. It is treated with *defibrillation* (electric shock that resets heart rhythm) that may be accompanied by epinephrine injections.

Review Questions

1. What is the pathway of blood flow through the pulmonary and systemic circulations?

2. Why is the myocardium of the left ventricle thicker than that of the right ventricle?

3. Describe the structure and function of the pericardium

4. What are the layers of the heart and their functions?

5. Describe the flow of blood through the heart, naming each chamber and the valves.

6. Describe the steps of the cardiac cycle and its pressures, volumes, and sounds.

7. Define EDV, ESV, stroke volume, cardiac output, and ejection fraction.

8. What is the cardiac pacemaker? Describe the conduction pathway through the heart.

9. Describe the conduction system of the heart and normal sinus rhythm.

10. What is the origin of each of the waves in an electrocardiogram?

Cardiac Muscle

The heart is a dual pump. The right half pumps deoxygenated blood through the lungs, where it becomes saturated with oxygen. The left half pumps that oxygenated blood to the entire body. One-way valves ensure that blood is pumped in the proper direction. Running this pump efficiently requires that contraction of the myocardium follows normal sinus rhythm: pacing begins in the sinoatrial node, passes over the atria, and then activates the AV node that sends pacing signals to the apex and back up through the ventricles.

Now we turn to considering how the amount of blood the heart pumps is regulated. At first, it seems a simple thing. If heart rate increases, the heart will pump more blood per minute. If the myocardium contracts with more force, a larger volume will be pumped with each beat. What we will seek to understand is how the heart changes its rate of beating and how the myocardium changes its strength of contraction. Increased cardiac output requires the heart to work harder. If oxygen supply is not sufficient to maintain the increased metabolic domain, pain (angina) may be felt in the chest.

Most clinical interventions with drugs that act on the heart reduce cardiac output which leads to lowered blood pressure and a heart that does not need to work so hard. In the next lecture, we will look into how changes in blood vessel diameter (peripheral resistance) influence blood pressure. Later, we study renal mechanisms that regulate blood volume, the third leg of the triumvirate (cardiac output, peripheral resistance, and blood volume) that determines blood pressure.

Cardiac Pacemakers

Explain the mechanism of spontaneous firing in cardiac pacemakers

Normal sinus rhythm begins in the sinoatrial node. The SA node sets the pace of the heartbeat. It is the *pacemaker* of the heart. The cells that make up the SA and AV nodes are modified muscle cells. They do not contract. They are *autorhythmic fibers* that generate action potentials spontaneously (over and over without any outside input).

Pacemaker cells differ from most cells by having a constantly rising membrane potential, the *pacemaker potential*. Rather than a steady resting membrane potential, at rest the membrane potential gradually becomes less negative. When this slow depolarization brings the membrane potential to *threshold*, an action potential fires. When the action potential finishes, membrane potential repolarizes and begins to rise again. The constantly rising pacemaker potential causes threshold to be reached over and over again to produce a rhythmic firing of action potentials. In the SA node, the pacemaker fires at a rate of about 100 bpm (every 600 msec).

Pacemaker cells of SA and AV nodes are autorhythmic
Pacemaker potential: slow depolarization of membrane potential
Action potential fires each time threshold is reached

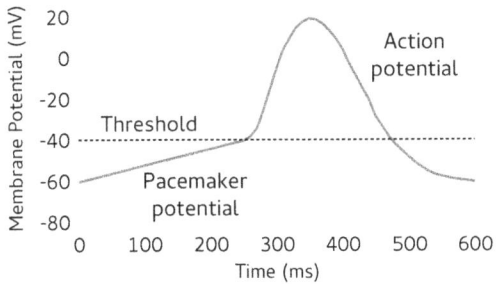

Why does the resting membrane potential of pacemaker cells drift upward? The pacemaker potential is also called *diastolic depolarization* as it occurs during diastole (when the heart chamber is filling with blood). It is initiated by influx of Na ions into the cells, which opens Ca channels, allowing an influx of Ca. Together the entry of these cations brings the membrane potential to threshold for opening of L-type (long-lasting) Ca channels and depolarization. Repolarization is due to closing of L-type channels.

Control of Heart Rate

Explain how the ANS changes heart rate by altering pacemaker potentials

Heart rate is controlled by parasympathetic and sympathetic innervation and circulating hormones. The parasympathetic *vagus nerve* slows the heart rate by decreasing the firing rate of the SA (sinoatrial) node. Increased heart rate up to about 100 bpm is mostly due to decreased vagal tone. Sympathetic innervation and circulating catecholamines (epinephrine) increase heart rate and the strength of cardiac contraction. Thus, parasympathetic innervation ("rest-and-digest") decreases cardiac output and sympathetic activation ("fight-or-flight") increases cardiac output. Maximal heart rate is 220 minus age in years (so 200 for a 20 year-old and 160 for a 60-year old).

Changes in heart rate are due to altered pacemaker potentials. Parasympathetic signals from the vagus nerve act on *muscarinic (M2) cholinergic receptors* to decrease the slope of the pacemaker potential. In addition, membrane potential dips deeper (becomes more hyperpolarized) before the pacemaker potential begins to rise again. Thus, it takes longer for the pacemaker potential to reach threshold.

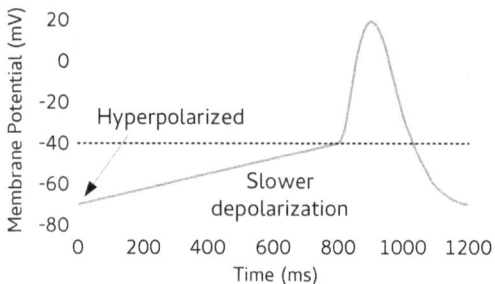

Sympathetic signals act through norepinephrine binding to *beta-1 adrenergic receptors*. The effect is opposite to that of the vagus nerve. Thus, sympathetic activation increases the slope of the pacemaker potential and the action potential begins sooner and is of shorter duration. The membrane potential does not need to become as negative before the pacemaker potential restarts its upward slope. Action potentials fire with greater frequency, leading to an increase in heart rate. Circulating catecholamines (epinephrine) have the same effect.

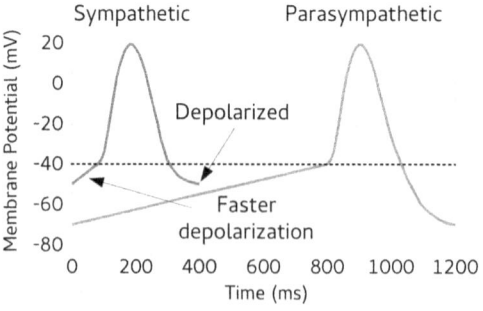

Cardiac Muscle Histology

Describe the unique features of cardiac muscle cells

Skeletal muscle are organized as long, multinucleated fibers. *Cardiac muscle cells* are nearly cuboidal, with single nuclei. Rather than forming side-by-side fibers, as in skeletal muscle, cardiac muscle cells individually branch to contact multiple nearby cells. They are connected to each other through specialized junctional complexes, the *intercalated discs*.

Intercalated discs are stabilized by two types of *anchoring junctions*, which join together the cytoskeleton of adjacent cells. *Desmosomes* (macula adherens) are spot junctions connecting intermediate filaments (keratin). *Fascia adherens* are ribbon-like junctions that join actin filaments (microfilaments) from one cell to another. The latter are similar in structure to zonula adherens in epithelial tissues.

> Highly branched cardiac myocytes are connected by intercalated discs
> Desmosomes and fascia adherens strengthen cell-cell interactions

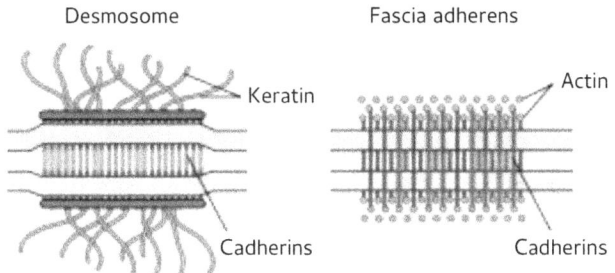

Within the intercalated discs are *gap junctions* that allow ions to pass from one cell to the next. Action potentials initiated in pacemaker cells cause action potentials in neighboring cardiac muscle cells. These muscle action potentials spread to nearby cardiomyocytes through gap junctions. The branching of cardiac myocytes ensures the spread of the electrical signal over the entire myocardium.

> Action potential in cardiac myocytes spreads out from cell to cell
> through gap junctions in intercalated discs of branching cardiomyocytes

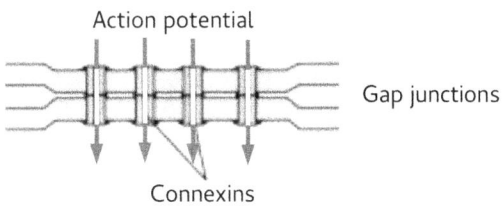

Action Potential Transmission in Cardiomyocytes

Explain how signals pass from cell to cell through gap junctions

The constantly rising pacemaker potential in the SA node causes threshold to be reached over and over again to produce a rhythmic firing of action potentials. Action potentials from the pacemaker cells spread to nearby cardiomyocytes that are connected together by gap junctions. Ions pass between cells through gap junctions to propagate the action potential throughout the myocardium. Pacemaker cells of the SA node spontaneously fire action potentials at a rate of about 100 per min. This rate drives the slower, 60 bpm natural firing of the AV node.

Gap junctions are connexin proteins that connect the cytoplasm of two adjacent cells, allowing small molecules and ions to pass between them. Gap junctions are located within intercalated discs, in the brain, and in smooth muscle and other cells. When voltage-gated Ca channels open in pacemaker cells, the resulting depolarization passes from the pacemaker cell to adjacent cardiomyocytes through gap junctions. Similarly, when cardiomyocytes depolarize, current (electrical energy) passes through gap junctions to the adjacent cell causing it to depolarize. By this means electrical current spreads throughout the myocardium.

The mechanism is similar to the flow of current in myelinated neurons, where depolarization at one node of Ranvier spreads through the myelinated region to the next node of Ranvier, which depolarizes, in turn. In cardiomyocytes, the current (electrical signal) of action potentials is transferred from one cell to another through gap junctions. Thus depolarization of one cell causes a current that travels to the next set of cells to which it is connected, each of which then depolarizes. The cells are said to be electrically coupled.

> in cardiomyocytes action potential current flows through gap junctions to depolarize the next cell to threshold to fire an action potential

Cardiomyocyte Action Potentials

Compare the properties of skeletal muscle and cardiac muscle action potentials

A typical heart rate is 70 bpm so action potentials will fire about every 800 msec and last some 300-400 msec. Thus, the cardiac muscle action potential is prolonged compared with that of skeletal muscle fibers. Cardiac muscle action potentials have a long plateau to match the slow contraction of cardiac muscle fibers. During the action potential, another action potential (and cardiac muscle contraction) cannot take place. This is the *effective refractory period*, which is much longer than that in skeletal muscle. It prevents summation. Unlike skeletal muscle fibers, tetanus is not possible in cardiomyocytes. All twitches are individual events.

> Skeletal muscle action potentials: short spikes that cause rapid contraction
> Cardiac muscle action potentials: prolonged and continue during contraction
> Long refractory period of cardiac AP prevents refiring before completion

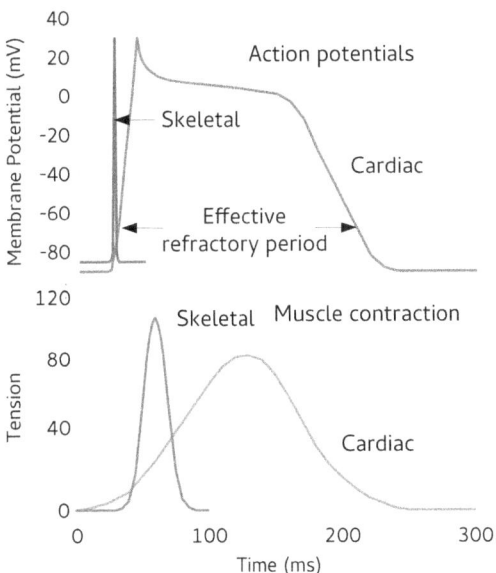

Cardiac Muscle Contraction

Explain the mechanism of cardiac muscle contraction

Cardiac muscle action potentials are divided into four phases. *Phase 0* is rapid depolarization. Fast voltage-gated Na channels open, just as in axons and skeletal muscle fibers. Opening is initiated by a small depolarization that is transmitted through gap junctions from adjacent cardiomyocytes.

Phase 1 is a slight repolarization due to the closing of these fast Na channels. *Phase 2* is the prolonged plateau phase characteristic of cardiac muscle cells. It is due to opening of slow L-type Ca channels and slow K ion channels. This prolonged plateau phase ensures that cardiac muscle contractions are slow and long-lasting, unlike the far briefer contractions in skeletal muscle that are initiated by a rapid neuronal action potential. *Phase 3* is repolarization, as L-type Ca channels close and rapid hERG (human ether a-go-go related gene) K channels open.

Phase 4 is the resting membrane potential (about -90 mV), that is due to open K leak channels and low Na permeability. Unlike the rising pacemaker potential of the SA and AV nodes, the resting membrane potential in cardiac myocytes is flat.

> Cardiac muscle action potentials are divided into 5 phases
> Prolonged plateau phase of cardiac muscle cells is due to
> L-type Ca channels that close slowly to extend refractory period

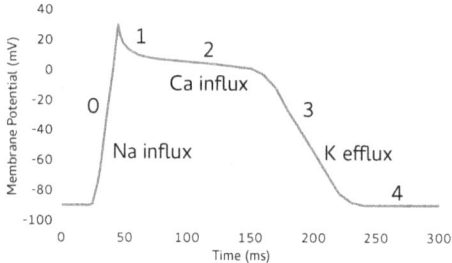

Ion Channels in Pacemakers

Describe the phases and ion fluxes in pacemaker action potentials

In pacemaker cells only phases 0, 3, and 4 are observed. In pacemaker cells, rapid depolarization in Phase 0 is due to influx of Ca rather than Na ions. Phase 4 drifts upward.

The upward drift of the pacemaker potential is initiated by influx of Na ions (*funny current*) through HCN (hyperpolarization-activated cyclic nucleotide-gated) channels. As the membrane depolarizes, T-type (transient) Ca channels open. Influx of Na and Ca ions brings the membrane potential to threshold for opening of L-type (long-lasting) Ca channels and depolarization (Phase 0). Phase 3 is repolarization due to closing of L-type channels.

> Cardiac pacemaker potential (phase 4) is due to Na funny current
> (HCN channels), then opening of T-type Ca channels to threshold
> L-type Ca channels open to depolarize (phase 0) then close (phase 3)

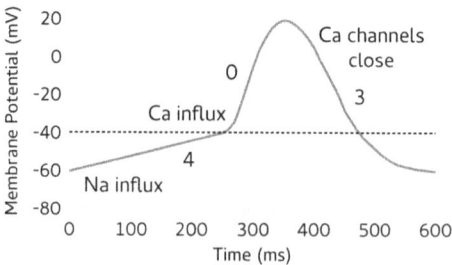

Drugs Acting on Cardiac Action Potentials

Explain how different classes of drugs alter cardiac action potentials

An arrhythmia is a disturbance of normal sinus rhythm. *Antiarrhythmics* are classified according to their primary mechanism of action. Class I drugs, such as *procainamide*, are Na channel blockers. Class II drugs (propranolol, metoprolol) are beta adrenergic receptor antagonists. Class III drugs (*amiodarone*) are K channel blockers. Class IV drugs (verapamil, diltiazem) are Ca channel blockers. All slow impulse conduction and generation.

Na and K channels are most important in cardiac myocyte action potentials, so Class I and III antiarrhythmics act on ventricular muscle fibers to treat life-threatening ventricular tachycardias. These slow ventricular impulse conduction and stop premature ventricular contractions (PVC). Class I drugs block fast Na channels, so do not act on pacemakers. They prolong the action potential and reduce conduction velocity in cardiomyocytes. Procainamide binds to open and inactivated fast sodium channels to prevent Na influx and slow the rapid upstroke of phase 0 in action potentials of cardiac muscle fibers. It prolongs action potential and refractory period which reduces impulse conduction velocity.

> Class I antiarrhythmics block fast voltage-gated Na channels
> Class II are beta-blockers; Class IV block L-type Ca channels
> Class III block voltage-gated K channels

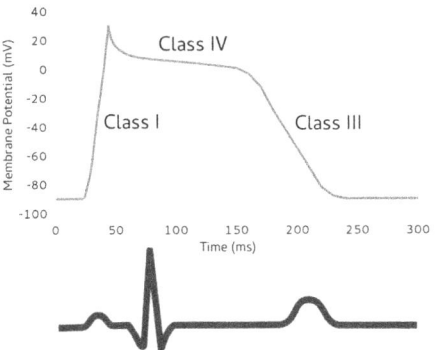

Class III drugs block K channels to slow ventricular repolarization. They prolong action potential duration and refractory period to prevent premature beats. Amiodarone is used to treat possibly fatal persistent ventricular tachycardia (VT) especially after cardioversion (restoration of normal sinus rhythm by drugs or electric shock). Amiodarone blocks potassium ion channels which delays repolarization of myocardial action potentials, increasing action potential and refractory period duration, preventing premature activation.

Beta-adrenergic receptors and Ca channels have effects on pacemakers, so Class II and IV antiarrhythmics act on SA and AV nodes and nodal conduction. In principle, they prevent supraventricular tachycardias, but these are usually not treated. Class IV drugs block open voltage-gated L-type Ca channels (calcium channel blockers = CCB). They prevent repolarization which prolongs effective refractory period in pacemaker cells. Propranolol and verapamil are most commonly used to decrease blood pressure.

Excitation-Contraction Coupling

Describe the source of calcium for contraction of cardiomyocytes

Na influx during phase 0 of the myocardial action potential opens L-type Ca channels. *Excitation-contraction coupling* describes the linkage between action potentials and release of calcium for contraction. L-type (DHP = dihydropyridine) Ca channels are found in all muscle cells. L-type channels slowly inactivate, so that the action potential of cardiomyocytes is prolonged while they are open. The influx of Ca through L-type channels triggers release of Ca from the sarcoplasmic reticulum through ryanodine channels (*calcium-induced calcium release*).

Influx of Ca through L-type (DHP) channels triggers release of Ca from SR
Ca spark: Ca released from SR that initiates contraction
Na pump builds Na gradient to drive out Ca ions with Na-Ca exchanger

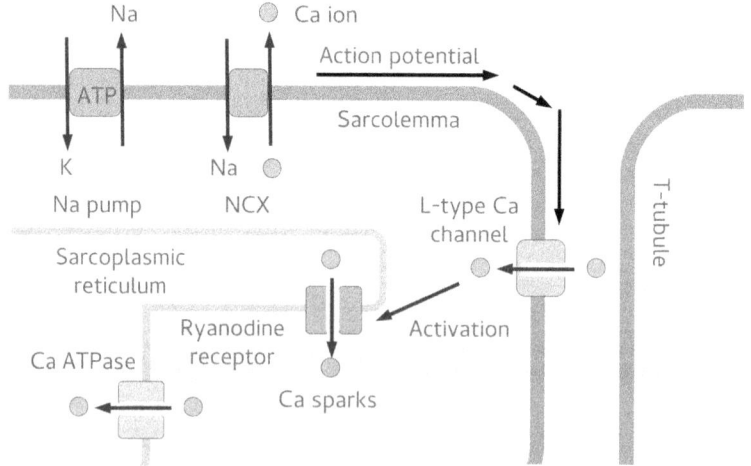

Unlike skeletal muscle fibers, DHP and ryanodine channels are not directly coupled in cardiac myocytes. Rather, cross-bridge formation between actin and myosin is initiated by a combination of extracellular Ca entry through L-type channels and Ca released from the SR. Ca balance in cardiac myocytes is maintained by coupled transporters. Na-K ATPase (Na pump) establishes a Na gradient that is used to drive Ca ions out of the cell via a *Na-Ca exchanger*.

Cardiac glycosides (derived from foxglove plants), such as *digoxin* inhibit Na-K ATPase so reduce extracellular Na available for Na-Ca exchanger. Increased intracellular Ca increases myocardial contractility (force of contraction) and prolongs SA node pacemaker potentials, which decreases heart rate. *Calcium channel blockers* (verapamil) prevent entry of Ca ions through L-type channels, decreasing Ca levels in contracting myocardial cells. This has the opposite effect to cardiac glycosides. The myocardium contracts less strongly, reducing stroke volume, decreasing cardiac output, and reducing blood pressure.

Myocardial Contractility

Explain how the strength of contraction of cardiomyocytes is regulated

Increased cardiac output is due to increased heart rate or stroke volume (volume pumped with each beat), or both. Heart rate is regulated by ANS effects on the pacemaker potential of the SA node. Stroke volume increases when cardiac muscle cells contract more forcefully. Unlike skeletal muscles, cardiac muscle cannot recruit more motor units or increase motor neuron firing rates to increase contractile strength. Rather, the force of contraction (*myocardial contractility*) is altered by changes in initial fiber length (preload) and inotropy, mechanisms independent of muscle fiber stretch.

Increased venous return increases end-diastolic volume (EDV) and *left ventricular end-diastolic pressure* (LVEDP), the ventricular pressure after atrial systole has completed. The increase in pressure stretches the myocardium in proportion to the increase in EDV. Since the increase in filling occurs prior to contraction, it is referred to as *preload*. When the myocardium is stretched, its force of contraction increases. Stroke volume then increases to match the increase in EDV, a relationship called the *Frank-Starling Law of the heart*.

These length-dependent changes are due to increased Ca affinity for troponin. The extent of actin-myosin filament overlap plays a only a minor role, as the length changes in the myocardium are very small compared with those in skeletal muscle fibers. By the Frank-Starling mechanism, the myocardium adapts its force of contraction to match stroke volume to venous return and ensure that the heart pumps the blood it receives.

> Preload: initial stretch of cardiac muscle (EDV) increases contractility
> Frank-Starling: force of contraction matches stroke volume to venous return
> Positive inotropy: increased myocardial contractility (beta-1 receptors)

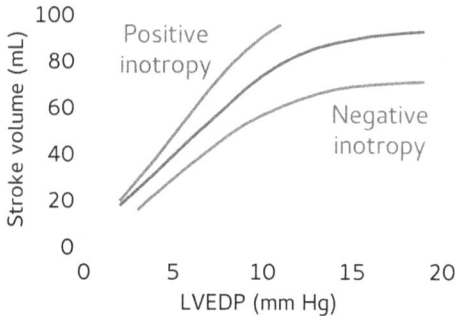

Positive inotropy is an increase in contractility at a given initial stretch of the myocardium. It is due to a combination of catecholamines (epinephrine and norepinephrine) acting on beta-1 receptors and decreased load against which the heart must pump (afterload). Positive inotropy is due to increased Ca concentrations and effectiveness. Ca influx in phase 2, Ca release by SR, and troponin-C binding to Ca all increase. Positive inotropy shifts the Starling curve upward: pumping is more effective at a given preload. Thus, sympathetic activation not only increases heart rate, it increases its force of contraction.

Heart Failure

Describe the compensatory mechanisms in heart failure and drugs used to treat it

Heart failure is the inability to maintain cardiac output sufficient to meet the demands of the body. Hypertension and coronary artery disease (CAD) are the most common causes. As the heart fails, it attempts to compensate to maintain blood flow and pressure. This compensation includes sympathetic activation, activation of the renin-angiotensin system that causes vasoconstriction and increases blood volume, and remodeling of the heart (myocardial hypertrophy). At first, compensation makes up for the reduced tissue perfusion by increasing cardiac output. However, myocardial remodeling eventually weakens or stiffens the heart, making it less effective in pumping of blood.

A common cause of heart failure is impaired coronary blood flow due to deposits in coronary arteries that narrow the blood vessel (*coronary artery disease*) or a clot (*myocardial infarction* = MI) that blocks flow and compromises oxygen delivery (*ischemia*). Valve disorders, congenital heart defects, and hypertension also increase workload and may damage the myocardium. Decreased contractility, often due to myocardial damage after MI, limits the ability of the heart to eject blood (systolic failure. Thickening of the ventricular wall due to remodeling leads to a stiffened (less compliant) heart wall, which fails to properly fill (diastolic dysfunction) and reduces stroke volume.

Decreased stroke volume due to failure of one side of the heart backs up blood on other side causing increased BP (pulmonary or systemic), congestion (edema), and failure on other side. For this reason, the disease is often called congestive heart failure. *Left-sided failure* is most commonly due to coronary artery disease (CAD) and hypertension. Blood backs up in the pulmonary system leading to *pulmonary edema* and shortness of breath (dyspnea). *Right-sided failure* is due to acute or chronic pulmonary disease. Blood backs up in the systemic circulation causing *peripheral edema* (ankles) and ascites (in abdomen).

Normal heart has adequate stroke volume with low end-diastolic pressures
Failing heart requires higher preload to maintain adequate stroke volume
Positive inotropic drugs partly restore adequate cardiac output

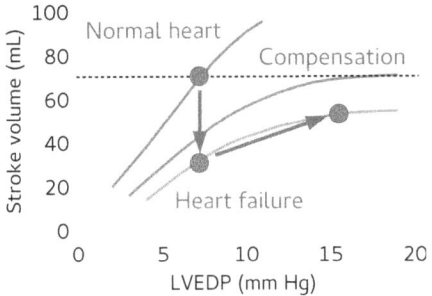

As the heart fails, it attempts to compensate to maintain blood flow and pressure, but is unable to do so. The heart muscle is weak, so more blood is left in the ventricles at the end of the cardiac cycle (EDV increases). This increases preload and the failing heart attempts to compensate by increasing contractility. It follows the Frank-Starling curve upward to restore adequate cardiac output. The increased EDV causes a decrease in ejection fraction (SV/EDV). Positive inotropic drugs shift the curve to a higher level, so that a lower preload is required for adequate cardiac output. Since EDV is decreased while SV is maintained, positive inotropy increases ejection fraction.

Cardiac glycosides (digoxin) have a positive inotropic effect due to increased intracellular calcium levels. Digoxin can be used in heart failure to increase myocardial contractility and make the heart work more efficiently. *Dobutamine* is a beta agonist that mimics the activity of the sympathetic nervous system (sympathomimetic), so it also is a positive inotrope. It must be given IV and is used in clinical settings to treat acute heart failure.

Beta blockers (metoprolol) are negative inotropes. They act by inhibiting the sympathetic response that compensates for a failing heart. Although compensation improves cardiac output for a short time, it also increases the workload on the heart. Beta-blockers decrease this workload and improve mortality. Calcium channel blockers (CCBs) that act on the heart (*verapamil*) have negative inotropic effects. They reduce intracellular calcium for muscle contraction. CCBs should not be used to treat heart failure.

Long-term treatment of chronic heart failure seeks to decrease workload on the heart. Diuretics increase urine volume which leads to a decrease in blood volume and pressure. Vasodilators open up blood vessels to reduce blood pressure. Beta blockers (*metoprolol*) decrease workload by inhibiting sympathetic activation.

Review Questions

1. What is unique about cardiac pacemakers that makes them fire over and over?

2. What is the effect of sympathetic and parasympathetic activation on pacemaker action potential?

3. How do changes in pacemaker action potentials change heart rate?

4. How do action potentials spread throughout the myocardium?

5. What role do desmosomes and fascia adherens play during contraction of the myocardium?

6. How do nerve impulses pass through gap junctions?

7. How are cardiac muscle action potentials and contractions different from those of skeletal muscles?

8. What are the phases of the cardiac muscle action potential? What channels account for each phase?

9. What is the source of Ca ions for myocardial contraction?

10. How does excitation-contraction coupling in cardiomyocytes differ from that in skeletal muscle fibers?

11. What is myocardial contractility? How does it change with increased venous return?

12. What is the effect of epinephrine and norepinephrine on myocardial contractility?

Blood Pressure

The heart pumps blood through vessels. These blood vessels deliver nutrients and oxygen to the tissues and remove waste products and carbon dioxide. They also ensure that carbon dioxide is exchanged for oxygen in the lungs. To do this, a sufficient volume of blood must be pumped through the system and at high enough pressure to ensure that it makes it through the entire circuit. Blood pressure is also important in pushing fluids out of capillaries to bathe the tissues with fresh nutrients and oxygen. At the same time, that blood pressure has to be low enough to allow the fluids to re-enter the circulation. So, the body must maintain blood pressure within narrow limits.

We studied how the rate and strength of contraction of the myocardium is influenced by input from the autonomic nervous system and hormones in the last lecture. We learned how the heart adapts its strength of contraction to the initial stretch of the walls of the ventricle, ensuring that blood which enters the heart is pumped out. This control of cardiac output is one of three components that determines blood pressure. Another is blood volume, which is largely regulated by the kidneys. We will study this regulation when we look at renal physiology and fluid balance.

In this lecture, we will investigate the tubes through which blood flows (blood vessels). By no means are these tubes rigid. Rather, they play an active role in regulation of blood pressure and the distribution of blood to ensure actively metabolizing tissues get sufficient blood flow and to regulate body temperature.

The two most important components of the blood vessel walls are their elastic fibers and smooth muscle cells. The elasticity of the walls of large arteries rebound after filling to keep blood flowing while the heart is filling (during diastole). Smooth muscle cells regulate the diameter of blood vessels. Most of this control is at the level of arterioles, which vasoconstrict to increase blood pressure and to divert blood from one tissue to another. Clinically, the most common problem in cardiovascular pathology is hypertension (high blood pressure). So we will need to dig deeply to understand regulation of smooth muscle contraction. In the next lecture, we will look at tissue perfusion, the supply of fluids to the tissues and their return to the circulation.

Blood Vessels

Describe the overall structure of blood vessels

Blood vessels have three distinct layers of tissues, or tunics. The *tunica intima* is the innermost layer of all blood vessels and the only layer in capillaries. It consists of simple squamous epithelium (*endothelium*) with an underlying basement membrane.

The *tunica media* is the middle layer of smooth muscle cells in blood vessels. It contains a varying amount of elastic fibers. The tunica media is thick in arteries and thin in veins. Arteries have an *internal elastic lamina* of elastic connective tissue between the intimal and medial tunics and an *external elastic lamina* between tunica media and externa. Contraction of smooth muscles cells in the tunica media narrows the blood vessel lumen (*vasoconstriction*). Relaxation of the tunica media expands the vessel lumen (*vasodilation*).

The *tunica externa* is the outer, fibrous connective tissue covering of blood vessels. In larger blood vessels, it contains arteries and veins, the *vasa vasorum* and nerves, the *nervi vasorum*. When the tunica intima blends into surrounding tissues in smaller vessels it called *adventitia*.

> Tunica intima: simple squamous endothelium and basement membrane
> Tunica media: smooth muscle cells; thick in arteries; thin in veins
> Tunica externa (adventitia): outer, fibrous connective tissue

Arteries

Describe the characteristic features of arteries

Arteries carry blood away from the heart. They are subdivided into the *elastic arteries* that are close to the heart, *muscular arteries* that distribute blood to the tissues, and the small *arterioles* that largely regulate blood pressure by altering their diameter. Elastic arteries include the major arteries, such as the aorta and its branches. Elastic arteries act like a rubber band. They expand when blood pressure rises during systole and recoil during diastole. This recoil pushes blood through the arterial tree to maintain blood flow when the heart is not contracting. So, the pressure of systole is due to contraction of the ventricles and the pressure of diastole is due to recoil of the greater arteries close to the heart.

Distribution, or muscular, arteries split off from the major arteries and travel to the tissues. Muscular arteries have smooth muscles cells that stretch and contract, damping the pulsatile flow caused by beating of the heart. As blood flows through arteries, blood pressure decreases, so that it is rather low (about 30 mm Hg) entering capillary beds.

Elastic arteries near heart expand with increased pressure of systole
and recoil to maintain blood flow during diastole
Muscular (distribution) arteries distribute blood to tissues

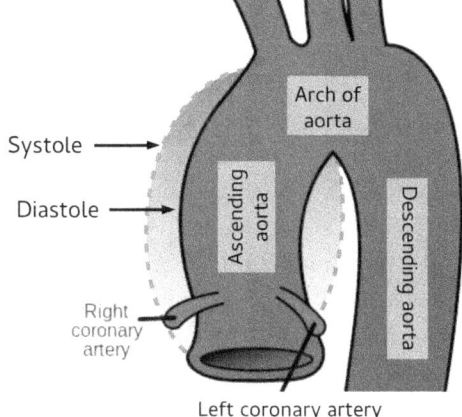

Veins

Describe the characteristic features of veins

Veins return blood to the heart. Their walls are thinner than those of the arteries and can expand to increase their volume, to a large extent. Blood pressure in the veins is low compared with that of the arteries. Thus, blood flow in the venous circulation needs some help to ensure that the blood returns to the heart.

Venous valves within the major veins of the appendages prevent backflow of blood in this low pressure system. They trap blood that might flow backward, ensuring it goes in one direction, much like the semilunar valves of the heart. This reduces the work of venous return of blood to the heart. Blood only needs to be pumped from valve to valve and not all the way from the extremities to the heart in one go.

Aiding venous return are the skeletal muscle and respiratory pumps. The *skeletal muscle pump* uses normal muscle contraction to assist in pumping blood (especially from the lower extremities) from valve to valve. For example, as the calf muscles contract while we move or wiggle our toes, they push blood upward toward the heart. In the thorax, inspiration lowers thoracic pressure, drawing blood into the heart. This is the *respiratory pump.*

Venous valves in major veins trap blood that flows backward
Skeletal muscle pump uses normal muscle contraction
to assist in pumping blood from valve to valve

Arterioles, Venules, and Capillaries

Describe the structure of small blood vessels

Distribution arteries give off *arterioles*, small arteries with smooth muscle cells that control the diameter of the blood vessels. Arterioles supply *capillary beds*, where nutrient and waste exchange occur. Changes in arteriole diameter regulate distribution of blood into capillary beds and resistance to blood flow through the arterial tree. Vasodilation of arterioles increases blood flow and decreases blood pressure. Vasoconstriction of arterioles decreases blood flow and increases blood pressure.

Capillaries have only a tunica intima, so that nutrients and wastes may readily exchange across the blood vessel wall. Capillaries are organized into networks, the capillary beds. Blood from capillary beds is collected by *venules*, small veins that have very thin walls. Even so, venules, like arterioles, have all three blood vessel tunics.

> Arterioles are small arteries with abundant smooth muscle cells
> Arterioles lead to capillary beds which are drained by venules
> Blood flow through capillary beds is regulated by vasoconstriction

Blood Pressure Measurement

Explain how blood pressure is measured in the brachial artery

Blood pressure is the force of blood that pushes against the walls of blood vessels. measured with a *sphygmomanometer*, an inflatable rubber cuff with a pressure gauge. Blood flow through the brachial artery is monitored by listening at the radial artery with a stethoscope. Cuff pressure applied to the upper arm stops blood flow in the brachial artery. Then, the pressure is gradually released.

> Korotkoff sounds: due to pulsatile, turbulent flow in radial artery
> As pressure is released, sounds are first heard when
> pressure is just below systolic and stop when below diastolic

No sounds are heard when blood flow stops (above systolic pressure) or is completely free to flow (below diastolic pressure). In-between these two pressures *Korotkoff sounds* are heard. These are due to pulsatile, turbulent flow in the radial artery. As pressure is released, sounds are first heard when the pressure is just below systolic, so this pressure reading is recorded as *systolic pressure*. As the pressure continues to be decreased, the point when no sounds are heard is recorded as *diastolic pressure*.

The blood pressure measurement is written as (systolic pressure)/(diastolic pressure). Normal BP = 120/80. *Pulse pressure* is the difference between systolic pressure and diastolic pressure. *Mean arterial pressure* (MAP) is the average blood pressure: MAP = diastolic pressure + 1/3 of pulse pressure. The new guidelines (2017) for hypertension by the ACC (American College of Cardiologists) and AHA (American Heart Association) is either systolic greater than 130 or diastolic greater than 80.

> Pulse pressure: difference between systolic and diastolic pressures
> Pulse is felt in surface arteries and damped by smaller, muscular arteries
> Mean arterial pressure = diastolic + 1/3 of pulse pressure

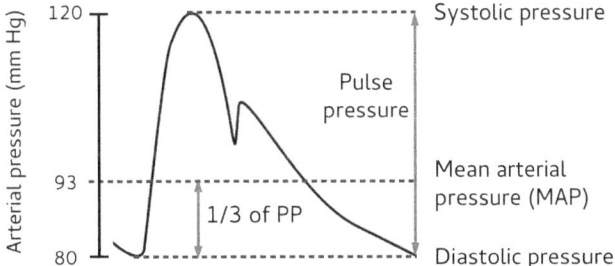

Determinants of Blood Pressure

Explain how blood pressure is influenced by cardiac output and vascular resistance

Blood pressure is determined by how hard the heart pushes blood through the circulation, the volume of blood that is pumped, and the resistance to its flow through blood vessels. Increasing heart rate increases blood flow through the body which increases blood pressure. Increasing the volume of blood pumped by the heart (stroke volume) also increases systemic blood flow and blood pressure. Together, heart rate and stroke volume determine cardiac output (CO = HR x SV). So, blood pressure can be regulated by regulating how fast (HR) and hard (SV) the heart contracts, both of which change *cardiac output.*

> Pressure is force of blood on vessel walls; increased blood volume
> pushes harder on walls of vessels and increases blood pressure
> Higher cardiac output pushes more blood through system so increases BP

Long term changes in *blood volume* also influence blood pressure. If water is retained by the kidneys (making a more concentrated urine), blood volume increases. Increased blood volume pushes against the walls of the blood vessels, directly increasing blood pressure. If blood volume is decreased (say, by the use of diuretics that increase urine volume), blood pressure decreases likewise.

At a given pressure, *vasoconstriction* decreases blood flow to downstream capillary beds, in proportion to the decrease in blood vessel area. To maintain the same blood flow requires an increase in blood pressure because it is harder to push fluids through a small tube than a larger one. We say the resistance to flow is higher in the smaller tube. This decrease in area is normally the result of contraction of arteriolar smooth muscle cells (vasoconstriction). It may also be due to occlusion by atherosclerotic plaque (buildup of cholesterol and fibrous tissue in arterial walls). Resistance to flow is proportional to the square of the area of the blood vessel. Only a small amount of vasoconstriction dramatically increases the blood pressure required to maintain flow.

Vasoconstriction or occlusion decreases blood vessel diameter and area
To maintain flow blood pressure must increase with vasoconstriction

| Area | 1.0 | 0.9 | 0.8 |
| MAP | 90 | 110 | 140 |

In the systemic circulation, resistance to blood flow through arterioles is *peripheral resistance*. The sum of all the resistance to flow in the systemic circulation is *systemic vascular resistance* (SVR). Since the resistance to flow in the pulmonary circuit is much less than that in the systemic circuit, SVR is nearly equal to total peripheral resistance (TPR).

Together, cardiac output and peripheral resistance are the two factors that determine blood pressure from moment to moment: BP = CO x SVR. They are not independent. Increased peripheral resistance increases afterload on the heart (how hard it has to pump), which reduces cardiac output. Vasodilation reduces afterload which increases cardiac output. So the two factors work together to maintain a relatively constant blood pressure.

Vascular Smooth Muscle Cells

Describe the innervation of vascular smooth muscle cells and mechanisms of vasoconstriction and vasodilation

The walls of arteries and veins, throughout the body, contain smooth muscle cells that regulate their diameter. They reside in the middle layer (tunica media) of the blood vessel wall. Smooth muscle cells surround *arterioles* (small arteries) at the entrance to capillary beds. Their diameter regulates blood flow and blood pressure. Norepinephrine released from varicosities in sympathetic motor neurons bathe smooth muscle cells. These *varicosities* are not distinct synapses like the neuromuscular junction, but are sites along the axon terminals that release neurotransmitter.

Sympathetic neurons release norepinephrine from varicosities
Neurotransmitter bathes the tissue to cause contraction
Gap junctions link smooth muscle cells together to coordinate contraction

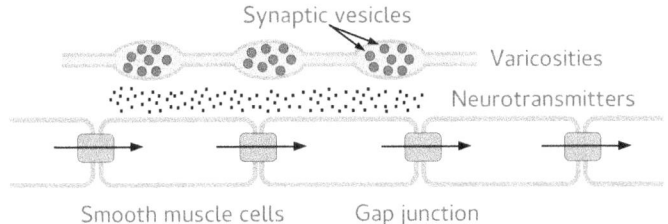

Gap junctions between smooth muscle cells allow signals to pass from one cell to another, allowing them to contract as a unit. Thus, activation of smooth muscle cells in the tunica media by sympathetic innervation causes *vasoconstriction* (decreased blood vessel diameter). *Vasodilation* is due to decreased sympathetic activity. Vascular smooth muscle cells are not innervated by the parasympathetic nervous system except in the clitoris and penis, where parasympathetic activation causes engorgement and erection by activating release of nitric oxide.

Smooth Muscle Cell Contraction

Describe the mechanism of contraction of smooth muscle cells

Smooth muscle cells are spindle-shaped. The contain both actin and myosin filaments. Actin filaments attach to dense bodies, which serve the same function as Z lines in skeletal and cardiac muscles. They do not have troponin. Rather, Ca binds to *calmodulin* which activates MLCK (*myosin light chain kinase*). When activated, MLCK phosphorylates myosin and initiating a cross-bridge cycle that pulls actin filaments toward the center of the cell. *Dense bodies* are attached through intermediate filaments to adherens junctions on the cell membrane. Thus, sliding of thin filaments by activated myosin pulls on dense bodies, causing the cell to contract.

Ca for contraction of smooth muscle cells comes from several sources. Depolarization of smooth muscle cell sarcolemma opens voltage-gated Ca channels that allow Ca to enter the cell. Ligand-gated Ca channels in the sarcolemma respond to neurotransmitters and hormones, causing additional Ca influx. *IP3-gated Ca release channels* in sarcoplasmic reticulum open in response to binding of hormones and neurotransmitters to G-protein coupled receptors on the cell surface, further increasing intracellular Ca. Relaxation is the result of declining Ca ion concentrations and activation of myosin light-chain phosphatase that inactivates myosin.

Ca binds to calmodulin which activates MLCK that phosphorylates myosin
Activated myosin pulls actin attached to dense bodies toward center of cell
Dense bodies attach through intermediate filaments to adherens junctions

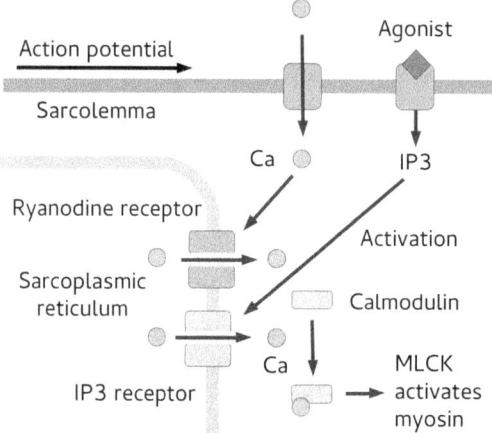

Calcium channel blockers (*nifedipine*) slow Ca entry into vascular smooth muscle cells. This opposes the increased Ca ion levels caused by sympathetic activation. Thus, CCBs inhibit vasoconstriction which lowers blood pressure and reduces coronary artery vasospasm.

ANS Receptors and Cardiac Output

Explain the effects of M2 and beta-1 adrenergic receptors on cardiac output

Sympathetic activation operates through binding of catecholamines (epinephrine and norepinephrine) to *adrenergic receptors*. Epinephrine is produced in response to sympathetic input to the adrenal medulla. It circulates throughout the body and acts as a hormone at sites distant from the adrenals. Norepinephrine is produced at sympathetic postganglionic axon terminals. It acts locally on smooth muscle tissues and in the myocardium. The response depends on the type of adrenergic receptor, of which three are critical to understanding regulation of cardiovascular function.

Norepinephrine and epinephrine both act on *beta-1 receptors* in the heart, which are located in the SA (and AV) node and in the myocardium. In the SA node, binding to beta-1 receptors increases heart rate. The effects are *chronotropic* (increased heart rate) and *dromotropic* (increased impulse conduction). In the myocardium, beta-1 receptors increase myocardial contractility. The effect is *ionotropic*. Thus, sympathetic activation increases cardiac output, which contributes to increased blood pressure.

The sympathetic effect on cardiac pacemaker activity is the result of norepinephrine and epinephrine binding to β-1 receptors that increase the slope of the pacemaker potential and cause depolarization. Increased myocardial contractility is the result of increased Ca influx from extracellular fluids, increased Ca release by SR, and increased Ca binding to troponin.

> Vagus nerve (parasympathetic) slows heart rate (muscarinic)
> Sympathetic cardiac nerves increase cardiac output by
> increasing heart rate and stroke volume (beta-1 adrenergic)

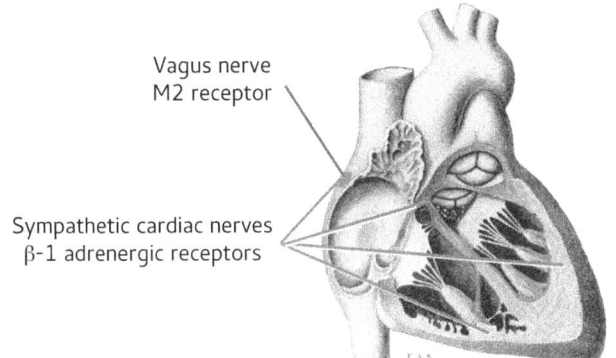

Vagus nerve
M2 receptor

Sympathetic cardiac nerves
β-1 adrenergic receptors

Adrenergic Receptors and Blood Vessel Diameter

Explain the effects of adrenergic receptors on blood vessel diameter

Smooth muscle cells in the vasculature have both alpha-1 and beta-2 adrenergic receptors. *Alpha-1 receptors* are widely distributed in skeletal muscles, skin, and the gastrointestinal tract. Alpha-1 (α-1) receptors in smooth muscle cells of the vasculature mediate vasoconstriction by stimulating Ca influx and release from SR. This vasoconstriction increases blood pressure. The abundant alpha-1 receptors in skin and GI tract also shift blood flow away from these sites in a fight-or-flight response. Parasympathetic activation decreases sympathetic tone, causing systemic vasodilation, but has no direct effect on the vasculature, except in erectile tissues (penis and clitoris).

> Sympathetic norepinephrine causes vasoconstriction (alpha-1 adrenergic)
> Vasodilation is due to decreased sympathetic norepinephrine or binding of
> epinephrine to beta-2 receptors of skeletal muscle in fight-or-flight

Sympathetic activation has more complex effects on skeletal muscle. In a fight-or-flight response, it is advantageous to increase blood flow to skeletal muscles. This is accomplished largely by activation of *beta-2 adrenergic receptors* which are abundant in skeletal muscle tissues. Unlike alpha-1 receptors, beta-2 receptors cause vasodilation. They are activated largely by circulating epinephrine and not so much by norepinephrine released by sympathetic neurons. Beta-2 receptors dominate the epinephrine effect, causing vasodilation in skeletal muscle and vasoconstriction in GI tract and skin. This diverts blood to muscle tissues in the fight-or-flight response.

Baroreceptor Reflex

Explain how the baroreceptor reflex regulates systemic blood pressure

The primary control center for blood pressure is in the medulla oblongata of the brain. The homeostatic response to changes in blood pressure is the *baroreceptor reflex. Baroreceptors* in the carotid sinus and aortic arch sense stretch in the blood vessel walls due to pressure and send signals to the medulla oblongata.

> Baroreceptors sense stretch in blood vessel walls
> Response to BP integrated in cardioregulatory center of medulla oblongata
> Sympathetic response increases BP by vasoconstriction and increased CO

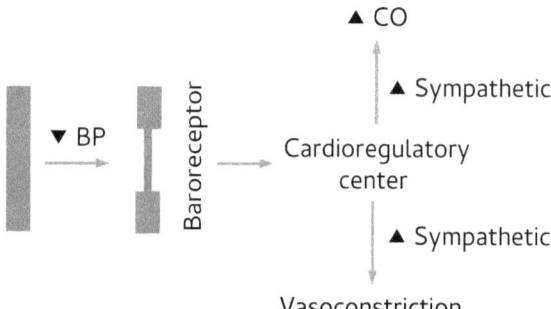

The *cardiovascular regulatory center* in the medulla oblongata responds to changes in blood pressure. With decreased blood pressure, it initiates a *sympathetic response*. Cardiac output is increased and the diameter of arterioles is decreased (vasoconstriction). Both of these work together to increase blood pressure. If blood pressure increases and becomes too high, a *parasympathetic response* through the vagus nerve decreases heart rate and sympathetic tone Smooth muscle cell contraction in arterioles is reduced, causing systemic vasodilation. Together, these responses reduce blood pressure back to its normal value.

Baroreceptors sense stretch in blood vessel walls
Increased BP integrated in cardioregulatory center of medulla oblongata
Parasympathetic response: vasodilation and decreased HR

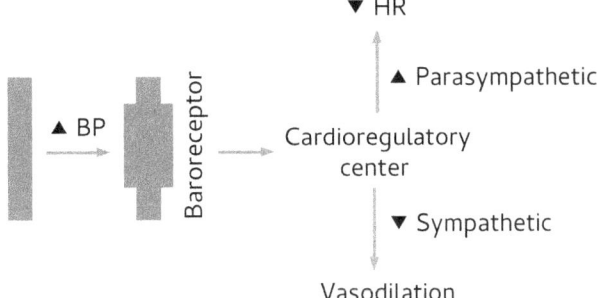

Cardiovascular Adrenergic Drugs

Explain how adrenergic drugs affect the cardiovascular system

Alpha-1 adrenergic blockers (*doxazosin*) prevent vasoconstriction in response to sympathetic activation. They cause peripheral vasodilation in gastrointestinal tract and skin. Doxazosin dilates arteries and veins to reduce blood pressure and relaxes smooth muscle cells around the prostate gland to reduce dysuria (pain or burning sensation with urination) due to urethral obstruction in BPH (benign prostate hypertrophy) or prostatitis (inflammation of the prostate).

Dobutamine is a beta-1 adrenergic agonist. It provides short-term inotropic support for cardiac decompensation in congestive heart failure. Its primary effect is to stimulate cardiac beta-1 adrenergic receptors to increase cardiac output. It also increases coronary blood flow and vasodilation through effects on beta-2 adrenergic receptors.

Dopamine is a nonselective adrenergic agonist (sympathomimetic) that activates both alpha and beta receptors to increase cardiac output and cause vasoconstriction. These combine to raise blood pressure in treatment of cardiogenic shock, an acute failure of the heart to pump enough blood at high enough pressure to perfuse tissues. *Epinephrine* similarly has nonselective adrenergic activity. Its beta-2 effects cause bronchodilation, so is indicated for anaphylactic shock.

Phenylephrine is a specific alpha1-adrenergic agonist that causes vasoconstriction in the nasal mucosa, so is effective as a short-term decongestant. In ophthalmic drops it induces mydriasis (dilation of pupils). It causes systemic vasoconstriction with no direct effect on heart, so may be used to treat circulatory shock.

Review Questions

1. What are the layers of the blood vessel wall?
2. How do the blood vessel wall layers differ in arteries and veins?
3. What is the importance of elastic arteries to maintaining blood pressure and flow?
4. How do elastic and muscular arteries differ?
5. What are venous valves and what is their function?
6. Distinguish between the skeletal muscle pump and respiratory pump.
7. Where are arterioles and venules located and how do they differ in structure?
8. Describe the structure of a capillary. Why is it important for capillaries to have thin walls?
9. How is contraction of smooth muscle cells in the tunica media of blood vessels controlled?
10. What role do gap junctions and varicosities play in smooth muscle cell contraction?
11. How does Ca activate smooth muscle cell contraction?
12. How does the organization of sliding filaments in smooth muscle cells differ from that of skeletal muscle fibers?
13. How do epinephrine and norepinephrine increase the strength of contraction of smooth muscle cells?
14. What are the effects of catecholamine binding to beta-1 receptors in the heart?
15. What role do adrenergic receptors play in controlling vasoconstriction?
16. Describe the baroreceptor reflex.

Tissue Perfusion

Pressure is required to push blood all the way through the circuits of the pulmonary and systemic circulations. It is also needed to push fluids out of capillary beds so that tissues may be bathed with fresh oxygen and nutrients. This is tissue perfusion. Waste products picked up by these fluids circulating around the tissues (interstitial fluids) are returned to the circulation either directly at the venous end of capillary beds or after passing through lymphatics, which return fluids to the veins near the heart. We learned in the last lecture that blood pressure is driven by cardiac output and depends on the diameter of blood vessels, particularly those of arterioles.

In addition to their role in regulating systemic blood pressure, arteriole vasodilation and vasoconstriction controls blood supply to the tissues. Sympathetic activation causes vasoconstriction of arterioles in skin and gastrointestinal tract, which not only increases systemic blood pressure, it reduces blood supply to these tissues in the expectation that we will not eat and run. Exercise causes vasodilation of skeletal muscle blood vessels which increases blood flow to these actively metabolizing tissues. While resting and taking in food, sympathetic tone is decreased and blood supply to the GI tract increases. Tissues also autoregulate blood flow to maintain perfusion when systemic blood pressure changes. This effect is particularly important in the brain.

Difficulties arise when flow through blood vessels is restricted by thickening of arterial walls by atherosclerotic plaque or blockage by blood clots. Both cause serious problems when they occur in the arteries of the heart (coronary artery disease) or in the carotid arteries which supply the brain (typically, where the external and internal carotids branch off from the common carotid at the carotid sinus). Clots may also form in veins (deep venous thrombosis) and travel to the lungs where they block blood flow or on damaged heart valves where they may be displaced and end up blocking arteries in the brain (stroke).

Not all the fluids that exit the capillaries return to the venous side of capillary beds. Those that do not are carried by the lymphatic system. In addition to interstitial fluids, lymphatics from the small intestine carry dietary fats that have been absorbed to the venous circulation, where they are seen first by the heart. While traveling through the lymphatics these fluids are exposed to specialized cells that protect the body from foreign invaders.

Coronary Artery Disease and Angina

Describe the effects of partial occlusion of coronary blood flow

The heart feeds itself first. The first branches of the aorta are left and right *coronary arteries*, which ensures that freshly oxygenated blood feeds the myocardium before passing on to the rest of the body. Any blockage of these arteries or the branches reduces blood flow to the heart and compromises its ability to function when under the stress of increased blood pressure or the demands of exercise. This is *coronary artery disease* (CAD). It most often results from partial occlusion of arteries due to accumulation of cholesterol and fibrous tissue in the intima (*atherosclerosis*). Decreased blood flow through a partially occluded artery is proportional to the decrease in blood vessel area.

> Partial occlusion of a coronary artery by atherosclerotic plaque decreases blood flow and compromises oxygen delivery to myocardium
> Blood clots in coronary arteries stop flow (myocardial infarction)

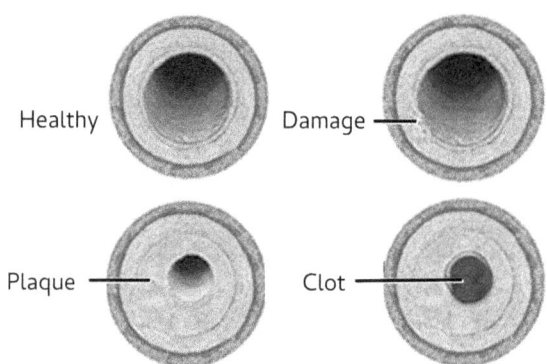

Angina is chest pain due to inadequate delivery of oxygen to the myocardium: oxygen demand by the myocardium exceeds oxygen supply by the coronary arteries. Hypoxia triggers release of adenosine and other mediators that activate cardiac pain receptors. Usually, angina pectoris is due to a combination of occlusion of coronary arteries by atherosclerotic plaques and increased blood pressure. Increased blood pressure increases the workload on the heart. Since this workload is during contraction it is called *afterload*. The combination of reduced blood supply and increased afterload is further increased by exercise and emotional stress, causing activation of pain receptors, which is angina.

Sites of damage to the arterial wall can cause blood clots to form. When a clot blocks a coronary artery, blood flow stops and the myocardium it normally supplies no longer receives oxygen (*ischemia*). This is a *myocardial infarction* (MI) or *heart attack*. An MI may also be caused by coronary artery spasms due to cocaine, stress, or cold. Infarction may so weaken cardiac muscle that the heart can no longer pump strongly enough to meet the metabolic demands of the body (*heart failure*). Damage to the myocardium releases *cardiac markers* into the blood that can be detected by clinical tests. *Troponin* is the most common marker. Its levels appear within 2-3 hours of injury and peak after 1-2 days. Occlusion can be evaluated by *coronary angiography*. A radiocontrast agent is injected into coronary arteries by coronary catheterization and blood flow visualized by x-ray images.

Blood Coagulation

Describe the stages of blood coagulation and common inhibitors

Hemostasis is the arrest of bleeding. It is divided into stages. In the vascular phase, endothelial cells contract (*vascular spasm*) to minimize blood loss. In the platelet phase, platelets attach to endothelial surfaces (*platelet adhesion*) to seal off the area. In the coagulation phase, a series of enzymes (*blood coagulation cascade*) are activated to form *thrombin* that converts soluble *fibrinogen* in the blood to an insoluble *fibrin* network. This fibrin network traps platelets and red blood cells to form a clot. Once a site heals, the clot retracts due to activation of circulating plasminogen to the enzyme plasmin which breaks down the fibrin network (*fibrinolysis*).

A *thrombus* is a blood clot that adheres to the vascular wall and may dislodge to become an embolus, which is a clot that circulates in the blood. These clots may lodge in and stop blood flow in a vessel. *Arterial thrombi* are platelet-rich and form at the site of endothelial lesions. *Venous thrombi* are fibrin-rich. In addition to lodging in coronary arteries, clots may block arteries supplying the brain, causing ischemic stroke. In a *transient ischemic attack* (TIA) blockage is temporary and usually causes no permanent damage.

Deep vein thrombosis (DVT) is the formation of clots in deep veins, typically in the lower leg or thigh. Here a clot can break loose to form an embolus that travels to right heart and enters the lungs. This is a pulmonary embolism, which blocks pulmonary arteries and is life-threatening. *Disseminated intravascular coagulation* (DIC) is a widespread activation of the clotting cascade due to sepsis or trauma. Clots form in small blood vessels throughout body and may lead to multiple organ failure.

Platelet aggregation inhibitors slow the activation and aggregation of platelets to prevent arterial thrombosis. Platelet activation releases ADP and thromboxane A2 that attracts and activates nearby platelets. *Aspirin* irreversibly inhibits the platelet COX-1 enzyme which is required to produce thromboxane A2. *Clopidogrel* is a pro-drug activated by first pass metabolism in the liver to a metabolite that irreversibly blocks platelet ADP receptors.

Anticoagulants (incorrectly called blood thinners) act on the blood coagulation cascade. They prevent and treat venous thrombosis (fibrin-rich clots) and systemic embolism. *Antithrombin III* is a natural enzyme that inhibits serine proteases, including thrombin and *Factor Xa* (which converts prothrombin to thrombin). *Rivaroxaban* is an oral drug that directly inhibits Factor Xa. *Heparin* binds to antithrombin III to increase its activity 1000-fold, thereby decreasing fibrin formation. *Vitamin K* is a required cofactor for postranslational modification of several clotting factors in liver. *Warfarin* inhibits *vitamin K epoxide* reductase that recycles epoxide back to vitamin K, reducing its levels and inhibiting formation of clotting factors.

Therapy with heparin and warfarin must be closely monitored to ensure blood coagulation is not inhibited so much that bleeding becomes uncontrolled. Heparin therapy is monitored by *aPTT* (activated partial thromboplastin time) in which clotting is activated by kaolin (clay). Warfarin therapy is monitored by *prothrombin time*. In this test, clotting is activated by tissue factor which is normally produced in response to tissue damage.

Thrombolytic agents dissolve existing clots by activating plasminogen to plasmin which cleaves fibrin. Alteplase is recombinant *tissue plasminogen activator* (tPA). It must be injected within 2-6 hrs of a thrombotic event (MI, acute ischemic stroke, pulmonary embolism) to prevent ischemic tissue damage.

Autoregulation of Blood Flow

Explain how blood flow changes in actively metabolizing tissues

Actively metabolizing tissues also produce metabolites that cause vasodilation to increase local blood flow (*active hyperemia*). These are *paracrines* because they act locally on tissues close to their site of release. Increased blood flow improves oxygenation and more effectively removes wastes.

Coronary blood flow is almost entirely dependent on local metabolic activity. Vasodilation in blood vessels of the myocardium is the result of increased metabolic demand during exercise and sympathetic activation. Myocardial blood flow is especially sensitive to brief periods of hypoxia (low oxygen supply) due to increased oxygen demand (active hyperemia). Compression of coronary arteries during systole causes a brief hypoxia which is made up by a reactive hyperemia that increases blood flow during diastole.

In the brain, the most important local vasodilator is carbon dioxide, which is associated with decreased pH (by formation of carbonic acid). Vasodilation in the *cerebral circulation* removes the excess carbon dioxide, maintaining pH balance.

Cutaneous Blood Flow

Describe the response of blood flow to changes in body temperature

Blood flow to the skin is largely determined by body needs for temperature control. When temperatures are high, blood flow to the skin increases for heat loss. Sweat glands are also activated to increase perspiration, which cools the skin as it evaporates.

When temperatures are low, blood flow to the skin decreases in a sympathetic response to maintain core temperature. In addition to cutaneous vasoconstriction, metabolic rate increases, and shivering may occur. In addition, blood flow is diverted to deep veins and, in extremes of cold, away from the appendages to protect core body temperature around vital organs. Exercise inhibits sympathetic tone controlling cutaneous blood flow, which increases heat dissipation by vasodilation of arterioles in the skin.

Nitric Oxide

Explain the mechanism of vasodilation by nitric oxide

Vascular endothelial cells, such as those in erectile tissue, release *nitric oxide* (NO) in response to parasympathetic stimulation. NO decreases intracellular calcium levels, causing MLCK dephosphorylation and smooth muscle cell relaxation. For this reason, NO is called *EDRF* (endothelium-derived relaxing factor) in the older literature.

Organic nitrates (*nitroprusside* and *nitroglycerin*) produce nitric oxide, locally. In the heart, they inhibit coronary vasoconstriction (spasm) and reduce preload. In the periphery, nitrodilators primarily increase compliance in small veins. Venous vasodilation decreases preload and myocardial work, reducing oxygen consumption by the heart.

Phosphodiesterase (PDE) inhibitors (*sildenafil*) increase vasodilation in the penis and treat erectile dysfunction. Inhibition of PDE prevents degradation of cGMP, the second messenger responsible for the effects of nitric oxide. Thus, nitric oxide normally released in response to parasympathetic stimulation of the penis maintains its activity (vasodilation that causes erection) for a longer time. PDE inhibitors also magnify the effects of organic nitrates on the heart and veins, causing severe hypotension. They should never be used together.

Tissue Perfusion

Explain how fluids are exchanged across capillaries and enter the lymphatic system.

Capillaries have the largest total cross-sectional area, so the slowest blood flow. They consist of a single layer of endothelial cells and are the site of transfer of fluids from the blood into the interstitial fluids. Lipid-soluble molecules, including oxygen and carbon dioxide, pass directly through cell membranes of endothelial cells. Water and salts pass between cells, but proteins do not readily pass through the endothelial barrier. The slow blood flow rate in capillaries ensures adequate exchange of nutrients and wastes with the tissues. This movement of fluids through the tissues is *perfusion*. Tissue perfusion maintains constant the environment surrounding cells and washes away any foreign material it may encounter.

Fluids are driven into the interstitial spaces (*filtration*) to bathe tissues in capillary beds by blood pressure. Since this pressure is due to water and occurs in capillaries, it is called *capillary hydrostatic pressure* (CHP). Pressure in interstitial fluids is nearly zero, so there is no direct resistance to the fluids being pushed out. They flow from a higher pressure (in the capillaries) toward a lower pressure (in the tissues).

So how do fluids return to the circulation? One way is through *lymphatic capillaries*, which have endothelial flaps that do not allow any fluids that enter to return to the tissues. But most of the interstitial fluids flow back into the capillary bed on the venous side (*reabsorption*). The driving force for this return cannot be interstitial fluid pressure because it is nearly zero. Some other force is required to bring fluids back into the circulation.

Consider the concentration of proteins in the blood vessels (which is high) and that in the interstitial fluids (very low). Recall that *osmosis* transports water from dilute solutions to more concentrated ones. That is, water will move in a direction that dilutes out high particle concentrations. The direction of osmosis is from interstitial fluids, with their low protein (particle) concentration to the blood, with its high protein concentration. Thus, osmosis pulls interstitial fluids back into the capillary bed. This pressure is *blood colloid osmotic pressure* (BCOP). It is *osmotic pressure* in the blood due to particles (colloids), which are primarily proteins. It is often called *oncotic pressure*.

Oncotic pressure is a rather low pressure, so it does not prevent fluids from flowing from the arteriolar side of a capillary bed into the tissue spaces. That is, blood pressure (CHP) is greater than the oncotic pressure of the blood (BCOP). The difference between the two, *net filtration pressure* (NFP = CHP − BCOP) is positive. Fluids flow out of the capillaries on the arteriolar side.

> Capillary hydrostatic pressure (BP) pushes fluids out of arteriolar capillaries
> Blood colloid osmotic pressure (oncotic pressure) is due to blood proteins
> Oncotic pressure pulls fluids back into venous capillaries where BP is lower

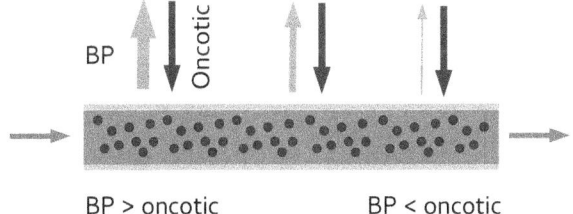

BP > oncotic   BP < oncotic

Blood pressure drops across a capillary bed to become significantly lower on the venous side. There, the oncotic pressure is greater than the blood pressure and fluids flow back into the capillaries. The net filtration pressure is negative (BCOP > CHP). This balance of blood pressure and oncotic pressure that determines the direction of flow of fluids between capillary beds and tissues are *Starling forces*.

Edema

Explain how fluids accumulate in tissues

Imbalances in tissue perfusion forces can cause accumulation of fluids in peripheral tissues beneath the skin or in cavities of the body (*edema*). In hypertension, increased blood pressure pushes fluids out of the capillaries and reduces its rate of return. However, increased venous pressure is a much stronger driver of *peripheral edema*. When the right ventricle fails, blood backs up in the systemic venous circulation, increasing capillary hydrostatic pressure which causes peripheral edema, especially in feet and legs. Increased blood volume due to activation of the renin-angiotensin system further increases pressure and edema. Fluid will sometimes accumulate in the abdominal cavity (*ascites fluid*).

Failure of the left ventricle increases back-pressure in the lungs leading to accumulation of fluid in the lungs (*pulmonary edema*). In *kwashiorkor*, acute protein malnutrition decreases plasma protein concentrations that determine oncotic pressure. Fluids do not return to capillaries and accumulate in tissues. *Lymphedema* is accumulation of fluids due to blockage of lymphatics.

Increased capillary permeability in *inflammation* may also cause local edema and swelling. *Histamine* and *bradykinin* released in response to tissue injury causes vasodilation of arterioles and vasoconstriction of venules which increases capillary hydrostatic pressure and contributes to inflammatory swelling.

Lymphatic Vessels

Describe the role of lymphatic vessels in circulation of body fluids

Interstitial fluids bathe tissues with nutrients and fresh oxygen and remove wastes and carbon dioxide. They also ensure cells are surrounded by a controlled environment of constant temperature, ionic composition, and pH. Most interstitial fluid returns to the circulation via capillaries. Some circulates back to the vascular system through *lymphatic vessels*.

> Interstitial fluids bathe cells then return to circulation via
> venous capillaries and lymphatic vessels (as lymph)
> Lymphatic capillaries: closed-ended with endothelial flaps that act as valves

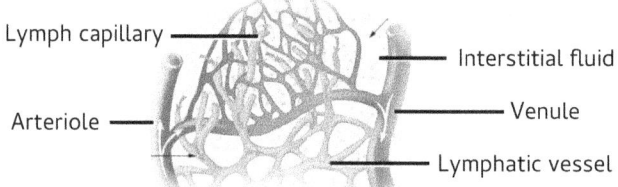

Lymph capillary — Interstitial fluid
Arteriole — Venule
Lymphatic vessel

Interstitial fluid enters lymphatic vessels through closed-ended lymphatic capillaries. *Lymphatic capillaries* consist of simple squamous epithelium and are highly permeable. They have flaps in their endothelium that act as valves to ensure one-way flow.

Once within the lymphatic system, the fluids are called *lymph*. This low pressure system with one-way flow driven by skeletal and respiratory pumps. Valves within lymphatic vessels prevent back-flow, much as they do in the venous side of the circulation.

Lymph trunks are large lymphatic vessels. They collect lymph from lymphatic vessels in the lumbar and intestinal regions. The cisterna chyli collects lymph from the lower trunks and contains newly absorbed lipids. It forms the thoracic duct. The thoracic duct drains the lower body and the upper left side of the body. The right lymphatic duct drains the upper right side of the body. Both collect fluids from lymphatic vessels and return them to the venous circulation through the subclavian veins.

Lymph Nodes

Describe the structure and function of lymph nodes

As lymph is carried back to the veins, it passes through *lymph nodes* that remove foreign invaders. Lymph nodes are enclosed in a fibrous capsule penetrated by a number of afferent lymph vessels that supply lymph to the organ. The fibrous capsule extends into the interior as trabeculae that support the overall structure. The lymph node is divided into functional compartments (*lymph nodules*) that develop a germinal center when challenged by a foreign molecule.

> Lymph nodes along lymphatic vessels remove foreign invaders before returning clean lymph to venous circulation

A reticular network of connective tissue provides a large surface area for adhesion of the primary cells of the immune response: dendritic cells, macrophages, and lymphocytes. Entering lymph passes by these cells which remove pathogens and activate the adaptive immune response that produces antibodies to further fight infection. Activated lymphocytes join the lymph which drains into the medullary sinus that empties into an efferent lymph vessel.

Review Questions

1. Where do the coronary arteries arise?
2. What is the relationship between blood flow and tissue perfusion?
3. What are the phases of blood coagulation?
4. How are clots removed?
5. How is blood flow autoregulated in metabolizing tissues?
6. How does cutaneous blood flow change to maintain body temperature constant?
7. How does nitric oxide cause vasodilation?
8. What are the pressures that drive fluids out of capillaries?
9. Why do fluids return to the venous side of capillary beds?
10. How do interstitial fluids return to the circulation?

11. What are the mechanisms of edema?
12. How do pulmonary and peripheral edema differ?
13. What are the special features of lymphatic capillaries that distinguish them from vascular capillaries?
14. How is lymph transported back to the venous circulation?
15. What is the role of lymph nodes?

Respiratory Physiology

An opening must be attempted in the trunk of the trachea, into which a tube or can should be put. You will then blow into this so that the lung may rise again. And the heart becomes strong.

– Vesalius

Ventilation

We now have a good understanding of how the circulatory system ensures that all of the tissues of the body are perfused with fluids. These fluids carry fresh oxygen and nutrients to the tissues and wastes away from them. In the next three sections, we consider how oxygen and carbon dioxide are exchanged in the lungs (*respiratory physiology*), how water-soluble wastes are removed from the blood (*renal physiology*), and how nutrients are digested, absorbed, and transported in the blood (*digestive physiology*).

Respiratory physiology begins with *ventilation*, the exchange of fresh for stale air in the lungs. This requires the lungs to expand on *inspiration*, which is an active process driven by expansion of the thoracic cavity to which the lungs are closely attached. Quiet breathing is mostly due to contraction of the diaphragm, which flattens to increase thoracic volume. Lungs are highly elastic, so their recoil to push air out when the diaphragm relaxes during *expiration*.

Once fresh air has entered the lungs, oxygen passes across the thin epithelium of tiny air sacs (*alveoli*) to enter plasma, then passes into red blood cells where it binds to *hemoglobin*. Binding to hemoglobin greatly increases the oxygen-carrying capacity of the blood. Carbon dioxide is primarily transported as the buffer *bicarbonate*. Bicarbonate enters red blood cells at the alveoli to be converted back to carbon dioxide that passes into the alveolar air and is expelled with expiration. The control center for ventilation is in the medulla oblongata which monitors *blood pH* to keep it nearly constant. When the blood becomes slightly acidic (pH decreases), ventilation increases to blow off more carbon dioxide and return the blood to its normal range.

Respiratory System

Describe the functional anatomy of the respiratory system

In the *respiratory system*, lungs exchange waste carbon dioxide of blood for fresh oxygen from the atmosphere. Anatomically, the system is divided into two tracts. The *upper respiratory tract* is in the head and neck. It extends from external nares (nostrils) through the larynx. It includes the nose, nasal cavity, paranasal sinuses, pharynx, and larynx.

The *lower respiratory tract* begins at the trachea and includes the bronchi and lungs.

> Respiratory system: passageways carry air to lungs
> for exchange of carbon dioxide in blood for fresh oxygen
> Upper respiratory tract: in head and neck; lower tract: within thorax

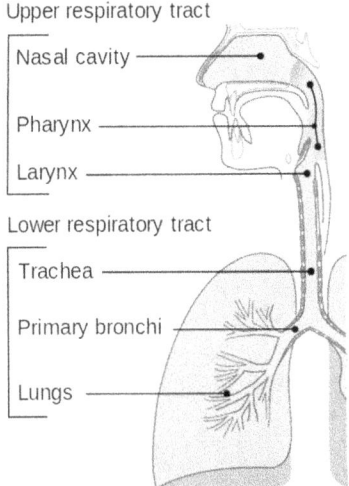

Upper respiratory tract
- Nasal cavity
- Pharynx
- Larynx

Lower respiratory tract
- Trachea
- Primary bronchi
- Lungs

Respiration is the exchange of oxygen and carbon dioxide between the atmosphere and the tissues. It has three components. *Ventilation* (breathing) exchanges gases between the atmosphere and the air sacs (alveoli) of the lungs. It occurs in the conducting zone of the respiratory system. *External respiration* is the exchange of fresh oxygen for waste carbon dioxide in the alveoli and occurs in the respiratory zone. *Internal respiration* is gas exchange between blood and tissues. The terms should not be confused with *cellular respiration* which is the use of oxygen by cells.

Upper Respiratory Tract

Describe the functional anatomy of the upper respiratory tract

The *conducting zone* of the respiratory tract has passageways that carry air to and from the exchange surfaces of the lungs. The airways filter, warm, and moisten the air. The *nasal cavity* is the open space above the hard palate of the mouth. Small hairs (vibrissae) within the nasal cavity trap and filter particles. Along the superior surface of the nasal cavity is the olfactory epithelium which detects odors. Air is warmed and moistened by the highly vascularized mucosa covering the turbinate bones (concha) of the nasal cavity.

Upper respiratory tract: conducting passages that bring air into trachea
Nasal cavity: hairs filter particles; mucosa warms and moistens air
Pharynx connects nasal and oral cavities to opening of larynx

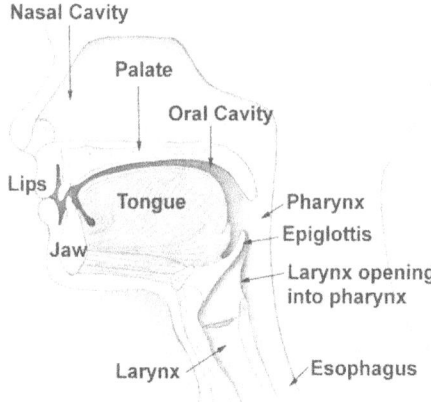

Around the nasal cavity are the *paranasal sinuses*. These air-filled spaces in the bones of skull (frontal, sphenoid, ethmoid, maxillary) are joined to the nasal cavity by small holes (ostia) which are readily blocked by inflammation or swelling. Blockage of ostia prevents drainage of mucus from the sinuses. Exchange of air between nasal cavity and paranasal sinuses is slow. Nasal cavity and sinuses modify the quality of sounds produced by vibration of the vocal cords.

The *pharynx* begins posterior to the nasal cavity and ends at the larynx. It is divided into three parts: nasopharynx, oropharynx, and laryngopharynx. The *nasopharynx* is behind the nasal cavity. The auditory (Eustachian) tubes link nasopharynx to the middle ear. The *oropharynx* is behind the oral cavity and is a passageway for air and food. The *laryngopharynx* connects the pharynx to the larynx.

The *larynx* is a triangular-shaped box containing vocal folds (cords) that vibrate to produce sound. The tension and length of the folds determines the pitch of the sound. Sound is modified by the mouth, tongue, pharynx, nasal cavities, and facial muscles. The vocal folds are open during inspiration. The *epiglottis* is a flap guarding the entrance to the glottis (opening) of the larynx. It rises upward while breathing and tilts downward during swallowing.

Larynx contains vocal folds; protected by cartilage
Epiglottis closes to block entrance to trachea (glottis) during swallowing

The larynx is protected by a set of cartilages. The most superior is the *thyroid cartilage* (Adam's apple), which is U-shaped and incomplete posteriorly. The cricoid cartilage lies inferior to the thyroid cartilage and forms a complete ring around the trachea. The thyroid gland is inferior to the thyroid cartilage. It surrounds the cricoid cartilage and laterally extends upward to the thyroid cartilage and downward to cover the first few tracheal cartilages. The arytenoid cartilage is the site of attachment of the vestibular folds (superior) and vocal folds (inferior). The vestibular folds help close off the glottis during swallowing.

Lower Respiratory Tract

Describe the functional anatomy of the lower respiratory tract

The lower respiratory tract extends from trachea to alveoli and includes the lungs. The trachea is anterior to the esophagus. It forms a straight tube down to the *carina*, where it branches into two *primary bronchi*. Tracheal cartilage is C-shaped to allow passage of food through the esophagus during swallowing, while preventing collapse of the tube. The trachealis muscle bridges the gap on its posterior side. These muscles constrict the trachea during coughing.

> Lower respiratory tract: conducts air to alveoli; includes lungs
> Trachea branches into two primary bronchi that enter lungs
> Bronchi divide to form small bronchioles that end in alveolar sacs

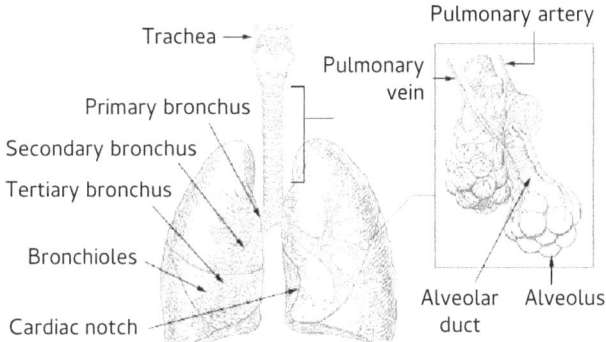

Primary bronchi, pulmonary arteries and veins enter the lungs at the hilum. The right primary bronchus divides into three secondary (lobar) bronchi that enter each lobe of the right lung. The left primary bronchus divides into two lobar bronchi that supply the two lobes of the left lung. The bronchi continue to branch into smaller and smaller divisions, the smallest of which are the *bronchioles*. At the level of bronchioles, the cartilage characteristic of bronchi is no longer present.

Bronchioles give rise to *alveolar ducts* that lead to alveolar sacs containing clusters of *alveoli,* where gas exchange takes place.

Mucociliary Escalator

Describe the process that removes inhaled particles from the airways

The respiratory passageways are lined with two different types of epithelium. The most anterior part of the nasal cavity (the vestibule) is lined with keratinized *stratified squamous epithelium,* similar to that of the skin. This epithelial lining transitions, above the palate and into the nasopharynx, to *ciliated pseudostratified columnar epithelia* characteristic of respiratory epithelium. In a pseudostratified epithelium, all of the cells make contact with the underlying basement membrane, but not all cells reach the apical surface. This epithelium gives the appearance of being stratified since the nuclei are not aligned, but it has only a single layer of cells.

> Goblet (mucous) cells secrete into mucus layer that traps particles in trachea
> Mucociliary escalator: cilia sweep trapped particles upward

The inferior parts of the pharynx are shared with the digestive system, so are exposed to food particles that would destroy the delicate respiratory epithelium. Thus, the oropharynx, laryngopharynx, and larynx are lined with stratified squamous epithelium.

The trachea and bronchi carry only air, so are lined with more fragile *respiratory epithelium*. Three cell types are characteristic of the respiratory epithelium: basal cells, goblet cells, and ciliated columnar cells. Basal cells are small, cuboidal cells that do not extend up to the apical surface. They renew the epithelium when it is damaged. *Goblet cells* secrete mucus that traps particles.

Ciliated columnar cells have motile *cilia.* A watery saline secretion covering the cilia allows the cilia to beat in unison. Together the mucous secretions and cilia make up the *mucociliary escalator* by which particles trapped in the mucus are carried upward by the beating cilia. This mucus is then swallowed or expectorated. When the cilia are damaged or excess mucus is produced, the normal sweeping motion is not sufficient and excess mucus must be coughed up.

Autonomic Regulation of Airways

Explain the effects of the ANS on airway diameter

Bronchoconstriction is narrowing of the bronchi and bronchioles. Bronchoconstriction causes dramatic increases in airway resistance that hinders the ability to breathe. Activation of the parasympathetic nervous system releases acetylcholine that causes smooth muscle con-traction by binding to muscarinic cholinergic receptors. This vagal input is the dominant neuronal pathway for control of airway smooth muscle tone. Sudden constriction of the airways (*bronchospasm*) is due to inflammatory mediators associated with asthma, chronic bronchitis, and anaphylaxis (severe allergic reactions).

Sympathetic innervation causes relaxation of smooth muscle cells in the airways by local release of norepinephrine from post-ganglionic axon terminals. Sympathetic activation of the adrenal medulla pro-duces mostly epinephrine and some norepinephrine that circulates through the blood to reach the smooth muscle cells of the airways. These catecholamines act on *beta-2 adrenergic receptors* to cause *bronchodilation* in the "fight-or-flight" response.

Asthma is a chronic pulmonary disease with inflammatory and bron-chospasm components. It causes breathlessness due to reduction in air flow through bronchioles (airway obstruction). Symptoms can be chronic and acute (commonly exercise-induced). *Beta-2 adrenergic agonists* relax bronchiolar smooth muscle, causing bronchodilation. They are divided into short-acting beta agonists (SABA) for acute at-tacks (*albuterol*) and long-acting beta agonists (*salmeterol*) that prevent bronchospasms.

Parasympathetic activity causes bronchoconstriction. *Anticholinergics* (*ipratropium*) reduce parasympathetic input. They relax airways to de-crease the work of breathing in *COPD* (chronic obstructive pulmonary disease, which includes chronic bronchitis and emphysema, often due to smoking). Anti-inflammatories (corticosteroids, such as inhaled *be-clomethasone*) provide long-term prevention of asthmatic attacks. Leukotriene receptor antagonists (*montelukast*) reduce effects of in-flammatory leukotrienes. Combination medications for asthma have both a LABA and corticosteroid.

> Bronchospasm: sudden narrowing of airways (parasympathetic)
> due to inflammatory mediators in asthma, chronic bronchitis, anaphylaxis
> Bronchodilation: beta-2 receptor agonists prevent bronchoconstriction

Cough, Sneeze, and Gag Reflexes

Explain the mechanisms of cough, sneeze, and gag reflexes

Bronchi and trachea are highly sensitive to chemical and mechanical irritation. Pulmonary irritant receptors located on the posterior wall of the trachea and pharynx, at the carina and entrance to the lungs, and in the diaphragm send signals to the *cough center* in the medulla oblongata. Cough can be an involuntary reflex or a voluntary response initiated by irritation and coordinated in the cerebral cortex. Motor neurons are sent from the cough center to the intercostals, diaphragm, and abdominal muscles. A cough consists of an inspiration, an expiration against a closed glottis, and a violent release of air when the glottis opens.

A *sneeze* is the result of irritation of receptors in the nasal mucosa. The soft palate and uvula depress while the tongue elevates to force most of the air through the nose and expel the irritant. Significant airflow through the mouth also occurs. Sneezing is triggered by histamine release from the mucosa, but may also be triggered by bright light in susceptible individuals. The reflex is integrated in the brainstem by input from the trigeminal nerves.

> Cough receptors send signals to medullary cough center for reflex cough
> Voluntary cough (cerebral cortex) is response to sense of irritation
> Motor neurons innervate diaphragm, intercostals, and abdominal muscles

Touching the posterior pharyngeal wall or soft palate causes a *gag reflex* (pharyngeal reflex). The soft palate is elevated and the pharyngeal muscles contract to seal off the pharynx and prevent entry of foreign materials into the larynx. Some individuals have a highly sensitive gag reflex that makes it difficult to swallow a pill.

Irritants may also cause *laryngospasm*, in which choking results from spasm of the vocal folds. Laryngospasm may result from inhalation of water in drowning and aspiration of vomit. It causes a feeling of suffocation and may result in hypoxia-induced loss of consciousness.

A *hiccup* is an involuntary, repeated contraction of the diaphragm followed shortly after by closure of the vocal cords, causing a distinctive 'hic' sound. Hiccups appear to serve no useful purpose.

Pleural Membrane

Describe the role of pleural fluid in breathing

The lungs are surrounded by a *pleural membrane*. This serous (double) membrane covers each lung separately so that damage to one lung sac does not damage the other. *Visceral pleura* directly covers the surface of the lungs and dips into the fissures between lobes. The *parietal pleura* is attached to the inner surface of the thoracic cavity (chest wall) and covers the upper surface of the diaphragm. The space between the pleural layers is the *pleural cavity*. It is filled with *pleural fluid* that allows slippage during breathing and keeps the lungs from collapsing.

Lungs surrounded by serous (double) membrane (pleura)
Parietal layer attaches to thoracic wall; visceral attaches to lungs
Between layers is pleural cavity filled with pleural fluid

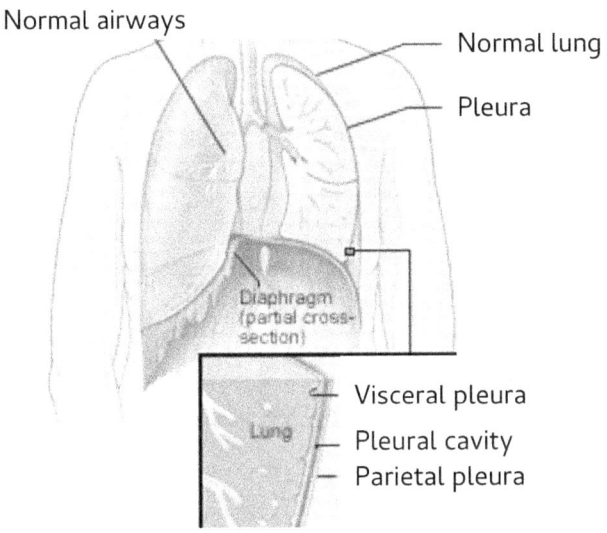

Normal airways

Normal lung

Pleura

Diaphragm (partial cross-section)

Lung

Visceral pleura

Pleural cavity

Parietal pleura

The lungs are highly elastic and constantly pull on the pleural fluid. The parietal pleura is tightly attached to the thoracic wall, so pulling by the lungs due to their elasticity causes a negative intrapleural pressure. Regardless of the stage of inspiration or expiration, the intrapleural pressure remains negative, keeping the lungs open. Disruption of the pleural sac (say, by a knife wound) breaks the connection between the lungs and thoracic wall and the elasticity of the lungs causes them to collapse (*pneumothorax*). During inspiration, expansion of the thorax pulls on the pleural fluid, causing the lungs to expand along with the chest wall. Pleurisy is inflammation of the pleura.

Lungs are elastic: they constantly pull inward on pleura to decrease intrapleural pressure that keeps lungs attached to thoracic wall
Damage to pleural sac may cause lungs to collapse (pneumothorax)

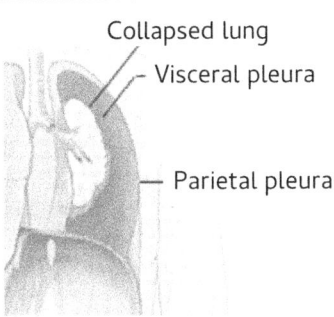

Collapsed lung

Visceral pleura

Parietal pleura

Inspiration and Expiration

Describe the pressure and volume changes during breathing

Pulmonary ventilation (breathing) exchanges gases between the atmosphere and air sacs (alveoli) of the lungs. *Inspiration* (inhalation) is active. It requires muscle contraction. Impulses from the phrenic nerve cause the dome-shaped diaphragm to contract and flatten. In addition, the ribs and sternum are elevated, expanding the rib cage. Additional muscles of the upper back and chest contract in *forced inspiration* to fill the lungs to a greater extent than does quiet breathing. *Quiet expiration* (exhalation) is passive. Muscles relax and the lungs recoil because they are elastic. *Forced expiration* enlists abdominal and intercostal muscles to push more air out of the lungs.

> Inspiration: diaphragm contacts and chest wall expands to enlarge thorax
> Decreased lung pressure causes air to enter lungs
> Expiration: muscles relax, pressure increases, air leaves lungs

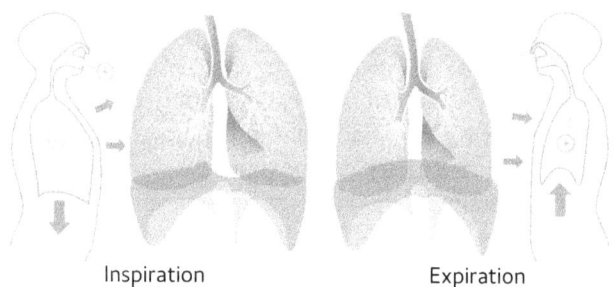

Inspiration   Expiration
pressure < atmospheric   pressure > atmospheric

> Boyle's Law: product of pressure and volume (PV) remains constant
> If volume increases, pressure decreases (inspiration)
> If volume is made smaller, pressure increases (expiration)

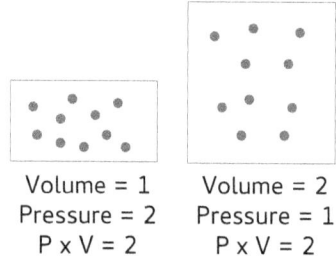

Volume = 1   Volume = 2
Pressure = 2   Pressure = 1
P x V = 2   P x V = 2

Inspiration and expiration follow *Boyle's Law*: as volume increases, pressure decreases. Thus, expansion of the lungs on inspiration increases lung volume, pressure becomes lower inside the lungs than outside and air flows into the lungs. Expiration decreases the volume of the lungs. Pressure increases slightly within the lungs and the air flows back out.

Lung Volumes

Define lung volumes and vital capacity

Lung volumes are measured by *spirometry*. This instrument measures the volume taken during a breath. *Tidal volume* is the amount of air moving into and out of lungs in each breath. It is about 500 mL during quiet breathing and can increase significantly with exercise. *Minute volume* is the volume breathed per minute, so it equals the tidal volume times the respiratory rate.

> Lung volumes are measured by spirometry; capacities are sums of volumes
> Tidal volume: quiet breathing; IRV: forced inspiration; ERV: forced expiration
> Vital capacity: maximum amount breathed (VC = TV + IRV + ERV)

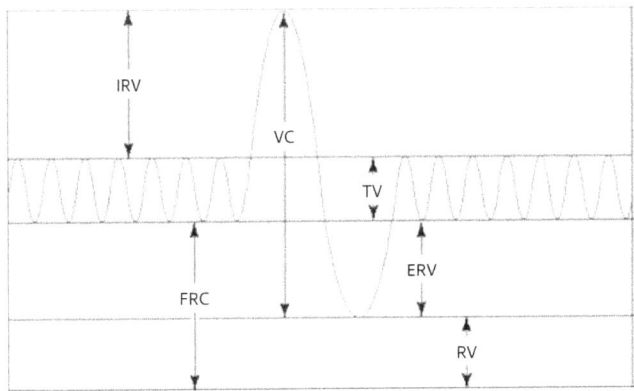

Inspiratory reserve volume (IRV) is the additional volume that can be inspired above that of quiet breathing. It is the result of forced inspiration. *Expiratory reserve volume* (ERV) is the additional volume of air that can be forced out the lungs (forced expiration). *Residual volume* (RV) is the volume of air that remains in the lungs and cannot be expelled. Lung capacities are sums of the individual volumes. *Vital capacity* is the sum of IRV + TV + ERV. Since this is identical to the amount of air that can be forced out of the lungs after a complete inhalation, it is usually called *forced vital capacity* (FVC).

Forced expiratory volume in 1 minute (*FEV1*) is the volume that can be blown out in 1 sec after a full inspiration. The ratio FVC / FEV1 should be 70-85%. *Peak expiratory flow* (PEF) measures the maximum speed of expiration and is easily measured with a hand-held flowmeter.

It varies depending on age, gender, and height. For a 25 year-old male 70 inches in height, PEF should be about 620 L/min.

Dead Space

Define dead space and how it is measured

Air taken into the lungs mixes with air that is already in the lungs. However, only the air in the respiratory zone can take part in gas exchange. Air in the conducting zone does not take part in gas exchange, so it is considered to be *dead space*. This anatomical dead space is a fixed volume of about 150 mL. With quiet breathing, anatomical dead space makes up a rather large fraction of the tidal volume, which is usually about 500 mL. This means that incoming air is significantly diluted by stale air.

Carbon dioxide concentrations in alveoli equal those in arterial blood (they equilibrate during gas exchange). If there were no dead space, exhaled carbon dioxide concentrations would be identical with arterial blood carbon dioxide levels. However, exhaled carbon dioxide is diluted by air that is not exchanged. Comparing exhaled carbon dioxide and arterial carbon dioxide concentrations gives a dilution factor that is used to measure dead space.

With deeper breathing, such as during exercise, tidal volume increases significantly but dead space does not change. Thus, the efficiency of ventilation improves because dead space is a smaller fraction of tidal volume. Breathing through a tube physically increases dead space and reduces the efficiency of ventilation. Endotracheal tubes inserted into airways do not dramatically increase dead space.

> Dead space: volume of inspired air not exchanged for
> fresh air during breathing (about 150 mL)
> Anatomical dead space: conducting zone and breathing tube volume

Pulmonary Compliance and Lung Surfactant

Define pulmonary compliance and explain the role of lung surfactant

The lungs are constantly pulling against the thoracic wall because they are elastic. Expanding the lungs during inspiration is somewhat like stretching a rubber band. It requires energy. This ability of the lungs to stretch is *compliance*. Stiff lungs have low compliance. They require more work to expand during inspiration. Lungs with low compliance are characteristic of *pulmonary fibrosis* in which damage to the lungs does not properly heal and normal tissue is replaced by stiff, fibrotic scar tissue.

Expiration is passive. The stretched lungs recoil because they contain abundant elastin fibers which act like a rubber band and snap back after being stretched. This elastic recoil is lost in COPD (*chronic obstructive pulmonary disease*) because of degradation of elastin. Such lungs have high compliance: they stretch easily but do not snap back. The lungs are easy to inflate but expiration must be forced to empty them. The end result is that a patient with COPD does not complete expiration and their lungs remain hyperinflated.

Lungs are also difficult to expand if they are filled with fluid. This occurs in pulmonary edema and in premature infants. Normally a thin layer of phospholipids and proteins coats alveoli to keep them from collapsing. In the absence of lung surfactant the moist surface of the lining of the alveoli is directly exposed to air. This air-water interface is difficult to stretch and tends to pull on the tissue to make the surface as small as possible. Lung surfactant acts as a detergent to reduce the surface tension that tends to cause alveoli to collapse. By this mechanism, the coating improves compliance and reduces the work required to expand the lungs.

Infants delivered before 34 weeks are at risk for *respiratory distress syndrome* (RDS) because their own lung surfactant is not yet being secreted in sufficient quantities. *Glucocorticoids* (*hydrocortisone*) given to the mother prior to birth speed up the developmental process and make RDS less likely. Maturity can be judged by amniocentesis. Infants with RDS are treated with CPAP (continuous positive airway pressure) using elevated oxygen levels. They may also be given a modified surfactant derived from cow lungs (*beractant*).

> Without lung surfactant, surface of alveoli are wet and hard to expand
> Lung surfactants coat alveoli, reducing surface tension
> making lungs easier to expand and less likely for alveoli to collapse

Respiratory Centers

Explain how rate and depth of ventilation is regulated

Quiet respiration is driven by an *inspiratory center* in the reticular formation of the medulla oblongata, a set of interconnected neurons extending through the brainstem. The synaptic connections (nuclei) of the inspiratory center form the *dorsal respiratory group* (DRG). Inspiration begins when neurons in the DRG increase their firing rate, causing contraction of the diaphragm via the *phrenic nerve*. Quiet expiration is passive, the result of relaxation of the diaphragm. The firing of the DRG follows a natural rhythm of nearly 2 sec of activity (inspiration) followed by 3 sec of inactivity (expiration) for a respiratory rate of 12-16 per minute.

> Inspiration is driven by inspiratory center in DRG of medulla oblongata which sets respiratory rhythm; forced ventilation is activated by VRG
> Pontine respiratory centers modify respiratory rhythm

A second respiratory center in the medulla is ventrolateral, so is named the *ventral respiratory group* (VRG). It has neurons that project to accessory muscles of breathing and regulates forced inspiration and expiration, in cooperation with the DRG. Within the VRG is the *pre-Bötzinger complex*, a nucleus essential for generation of the respiratory rhythm. It contains pacemaker cells, but the precise mechanism of rhythm generation is unknown.

Two pontine nuclei (located in the pons, just above the medulla oblongata) modulate breathing. The *apneustic center* sends signals to the DRG to prolong inspiration by delaying the shut-off of signals to the diaphragm. It is inhibited by *pulmonary stretch receptors* that prevent overstretching of the lungs by a too-prolonged inspiration. The *pneumotaxic center* is a higher center in the pons that inhibits the apneustic center to regulate the depth and rate of breathing in response to body needs. Higher brain centers can voluntarily over-ride the automatic rhythm of breathing when we are conscious.

Regulation of Ventilation

Describe how ventilation is regulated in the brainstem

The rate and depth of breathing is regulated primarily by carbon dioxide levels. Poor ventilation leads to increases in blood carbon dioxide and decreased blood pH. The primary chemoreceptors that monitor carbon dioxide levels in the blood are located in the medulla oblongata. There, carbon dioxide passes across the blood-brain barrier.

> Central chemoreceptors in medulla oblongata respond to pH
> Death and rate of respiration increase in response to increased pH
> Carbon dioxide passes through BBB and reacts with tissue carbonic acid

In cells near *central chemoreceptors*, carbon dioxide is converted by *carbonic anhydrase* to carbonic acid. Carbonic acid dissociates back to carbon dioxide and hydrogen ions that are detected by the receptors. Thus, central chemoreceptors respond to local pH changes in CSF and indirectly to blood pH. Signals from central chemoreceptors activate brainstem respiratory centers to increase the rate and depth of breathing when blood carbon dioxide levels increase.

Review Questions

1. What are the functions of the respiratory system?
2. How do the upper and lower respiratory tracts differ anatomically?
3. What are the functions and major structures of the conducting zone?
4. How do bronchi differ from bronchioles?
5. Describe the structure and functions of the mucociliary escalator.
6. What causes bronchospasm and bronchodilation?
7. What is the mechanism of cough?
8. What is the structure and function of the pleural cavity and fluid?
9. What holds the lungs attached to the thoracic wall? How does this explain lung collapse in pneumothorax?
10. How does contraction of the diaphragm change thoracic volume and pressure in ventilation?
11. How does Boyle's law explain the mechanism of ventilation?
12. What are the various lung volumes and how is vital capacity defined?
13. What is dead space?
14. What is compliance and how does it change in disease states?
15. What is the function of lung surfactant?
16. Where are the regulatory centers for ventilation located? How do they differ?
17. How is ventilation regulated by central chemoreceptors?

Gas Exchange and pH Balance

Ventilation is responsible for ensuring that stale air in the lungs is rapidly exchanged for fresh air in the environment. Once fresh air enters the lungs, carbon dioxide and oxygen exchange across the respiratory epithelium of the alveoli to reach equilibrium between alveolar air and gases dissolved in the plasma (*gas exchange*).

The solubility of oxygen in plasma is not high enough to supply tissues with adequate amounts to maintain metabolism. For this reason, oxygen binds to *hemoglobin* within red blood cells to increase its concentrations in blood. Carbon dioxide also enters red blood cells to be converted to *bicarbonate* which circulates in the blood and contributes to *pH balance*. Increased ventilation gets rid of carbon dioxide and keeps blood from becoming too acidic.

Alveolar Epithelium

Describe the histological structure of the alveolus

Bronchioles branch until they form *respiratory bronchioles*. The walls of the smallest airways gradually thin to simple simple cuboidal epithelium in the terminal and respiratory bronchioles. Beginning at the respiratory bronchioles, the epithelium is thin enough and the capillary bed extensive enough for gas exchange. From the respiratory bronchioles, small tubes (alveolar ducts) lead to *alveolar sacs* containing multiple *alveoli,* which are the blind-ended pouches. Together, these structures form the *respiratory unit* where gas exchange is possible. Alveoli are lined by a very thin *simple squamous epithelia* and are highly vascularized. Nearly all of the gas exchange takes place in the alveoli.

The *alveolar epithelium* consists of three cell types. *Type I cells* are thin, flat cells (simple squamous epithelium) that line the alveoli. *Type II cells* produce lung surfactant starting at 32 weeks of gestation. Lung surfactant coats the interior surfaces of the alveoli, reducing surface tension. This keeps the alveoli from collapsing at end-expiration. *Alveolar macrophages* (dust cells) engulf particles that make it into the alveoli.

Bronchi branch to form bronchioles that have no cartilage
At ends of smallest bronchioles, alveolar ducts lead to alveolar sacs
Alveoli are terminal structures where gas exchange takes place

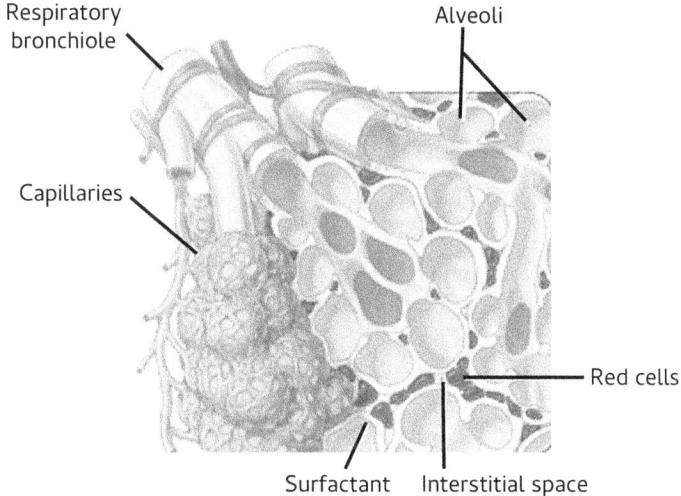

Respiratory bronchiole

Alveoli

Capillaries

Red cells

Surfactant   Interstitial space

The *respiratory membrane* consists of a Type I cell, capillary endothelial cell, and their fused basement membranes. It is the site of gas exchange in the alveoli. The respiratory membrane is very thin (about 2 um thick), so gases rapidly pass through by diffusion and come to equilibrium between the air of the alveoli and the plasma in the surrounding capillaries. Low blood pressure (capillary hydrostatic pressure) in the lungs minimizes flow of fluids out of the pulmonary capillaries into the limited interstitial space of the alveoli.

Respiratory epithelium: type I alveolar epithelial cell, capillary
endothelial cell, and fused basement membranes (very thin)
Gases diffuse through epithelium to reach equilibrium with plasma

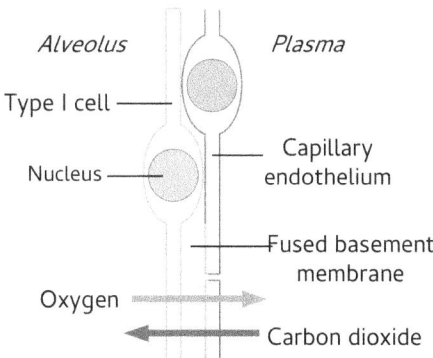

Alveolus   *Plasma*

Type I cell

Capillary endothelium

Nucleus

Fused basement membrane

Oxygen

Carbon dioxide

Partial Pressures

Define partial pressure and states its significance

Three gas laws are important for understanding pulmonary ventilation. Each is named after its discoverer: Robert Boyle, John Dalton, and William Henry. *Boyle's Law* determines the direction of gas flow in ventilation: as volume increases (inspiration), pressure decreases and gas flows into the lungs.

According to *Dalton's Law*, the total pressure of a gas in a mixture of gases is the sum of the pressures of the gases if they were present alone. In air, the concentration of oxygen in the air is 21%. At sea level, air pressure is 760 mm Hg. Thus, the *partial pressure* of oxygen in air is 21% of 760 mm Hg or P_{O_2} = 160 mm Hg. Most of the rest of air is nitrogen.

> Partial pressure: pressure of an individual gas in a mixture of gases
> Dalton's Law: total pressure equals sum of partial pressures
> Partial pressures are used for gas concentrations in liquids and gases

Dalton's Law applies to gases in air and gases dissolved in blood plasma. If pure air is brought into contact with plasma and allowed to equilibrate, the concentration of oxygen in the air and in plasma would become identical: P_{O_2} in air and plasma would be 160 mm Hg. This is *Henry's Law*: the amount of dissolved gas is proportional to its partial pressure in the gas phase. Thus, partial pressures can be used to measure the concentration of a gas in air and in blood plasma.

> Henry's Law: at equilibrium, pressure of a gas in liquid equals its
> pressure in air above liquid; gases diffuse down partial pressure gradient
> Solubility of oxygen in air is much higher than in plasma

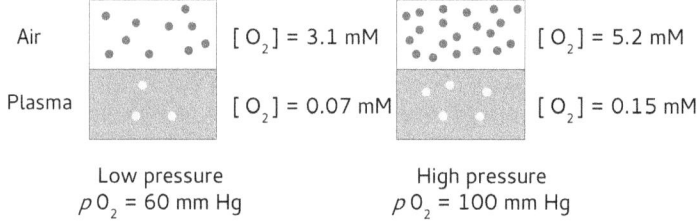

Air $[O_2] = 3.1$ mM $[O_2] = 5.2$ mM

Plasma $[O_2] = 0.07$ mM $[O_2] = 0.15$ mM

Low pressure   High pressure
$pO_2 = 60$ mm Hg $pO_2 = 100$ mm Hg

Like all types of diffusion, the direction of gas diffusion between air and plasma is toward equilibrium, where the concentrations become identical. The direction of diffusion of oxygen is from high partial pressures in air toward low partial pressures in plasma entering alveolar capillaries. Conversely, the direction of diffusion of carbon dioxide is from high partial pressure in entering blood toward the lower partial pressure of air in alveoli. At the alveoli, oxygen enters plasma and carbon dioxide leaves it.

Partial pressures express the concentration of a gas using the same terms in a mixture of gases as for gases in solution. Since the solubility of oxygen in water is low, the *amount* of oxygen dissolved in plasma is much less than it is in air, even when the partial pressures are identical. The direction of gas diffusion is determined by its partial pressure. Gases diffuse from high partial pressures to lower partial pressures.

Alveolar Gas Exchange

Explain how to predict the direction of gas diffusion

Nature tends toward equilibrium. At equilibrium, according to Henry's law, the partial pressures of a gas above a liquid and within the liquid are equal. Thus, the direction of diffusion is from higher partial pressure toward lower partial pressure. That is, the direction of diffusion tends to equalize the partial pressures in the gas of the alveoli with those of the incoming plasma.

Typically, the partial pressure of oxygen entering the capillary beds surrounding the alveoli is about 40 mm Hg and that in the alveolar air space is 100 mm Hg. The alveolar air has mixed with stale air of the conducting passages of the pulmonary airways, so the partial pressure of oxygen is significantly lower than that in the atmosphere. In spite of this dilution, the partial pressure of oxygen remains higher in the alveolar air than in the nearby plasma, so oxygen diffuses into the plasma. Air and plasma leaving the alveoli has an oxygen partial pressure of about 100 mm Hg. Conversely, carbon dioxide partial pressures entering alveolar capillaries is about 45 mm Hg, while that in the alveolar air is 40 mm Hg. Carbon dioxide concentrations also come to equilibrium so that plasma leaving the alveoli has a decreased, but still significant carbon dioxide concentration of 40 mm Hg.

> Blood entering lungs has high carbon dioxide and low oxygen concentrations
> In alveoli, plasma reaches equilibrium with alveoli air due to
> diffusion of carbon dioxide out of plasma and of oxygen into plasma

Ventilation and blood flow through the alveoli are generally matched so that alveolar gases come to equilibrium with gases dissolved in the plasma of the surrounding capillaries. That is, blood flow is slower than gas exchange across the thin respiratory membrane. Gas exchange may be limited in disease states due to a thickened alveolar membrane or edema.

Oxygen Transport

Explain the mechanism of oxygen transport in blood

At the alveoli, oxygen passes across the respiratory membrane by diffusion to enter plasma. However, oxygen is not very soluble in plasma. To transport sufficient oxygen to meet the needs of the body, it must enter red blood cells and bind to *hemoglobin*, a red pigment which carries some 98% of oxygen in the blood. This allows high concentrations of oxygen to be transported, in spite of its low solubility. Binding to hemoglobin pulls yet more oxygen into red blood cells as the plasma partial pressure of oxygen equilibrates with that in the alveoli.

At the tissues, oxygen is released from hemoglobin, passes through the plasma, crosses endothelial cells, diffuses through interstitial fluids and enters cells where it is used for aerobic metabolism. Oxygen uptake and aerobic metabolism within cells pulls more oxygen into the cells and further depletes hemoglobin of its oxygen. Dissociation of oxygen from hemoglobin is further increased with elevated temperature, increased local carbon dioxide concentrations, and lower pH associated with rapidly metabolizing tissues.

Alveolar oxygen enters plasma and diffuses into red blood cells
Most oxygen (98%) is transported in blood bound to hemoglobin (Hb)
Oxygen released from Hb at tissues passes through ECF and enters cells

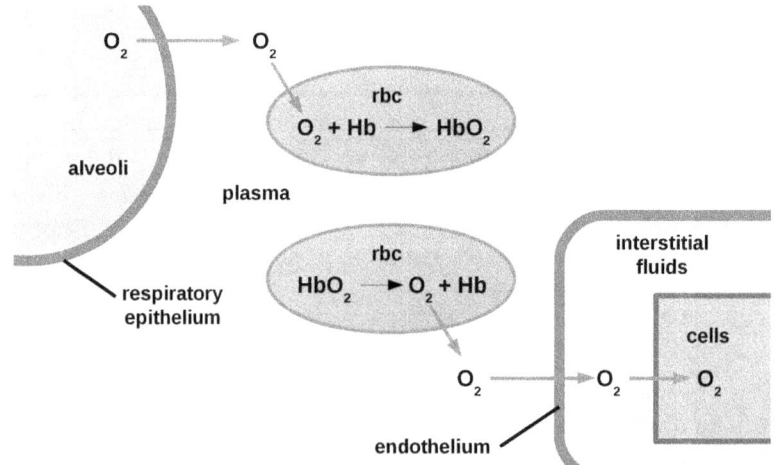

Carbon Dioxide Transport

Explain the mechanism of carbon dioxide transport in blood

Like oxygen, carbon dioxide is highly soluble in cell membranes. Both pass readily by simple diffusion across the respiratory epithelium, red blood cell membranes, endothelial cells lining blood vessels, and in and out of cells in tissues. However, carbon dioxide is more soluble in plasma. About 8% of carbon dioxide is transported directly in plasma. The remainder enters red blood cells where some binds to hemoglobin (about 23%) and the rest is converted to *bicarbonate*. The enzyme *carbonic anhydrase* catalyzes addition of water to carbon dioxide to form carbonic acid, which dissociates to form hydrogen ions and bicarbonate. Bicarbonate is pumped into plasma in exchange for chloride ions.

The result is most carbon dioxide transport is in the form of bicarbonate in plasma. At the alveoli, bicarbonate re-enters red blood cells in exchange for chloride ions. There it is converted back to carbon dioxide by carbonic anhydrase. Hemoglobin also releases its store of carbon dioxide. These leave the red cell, diffuse through plasma, and cross the respiratory epithelium to enter alveolar air.

> Tissue carbon dioxide leaves cells and passes through ECF to plasma
> Carbon dioxide in red cells binds to Hb or is converted to bicarbonate
> Bicarbonate enters plasma for transport and returns to RBC in lungs

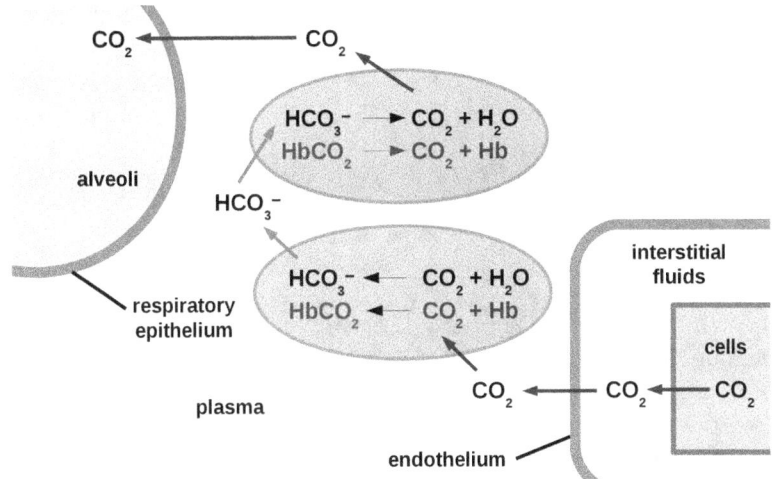

Bicarbonate Buffer

Explain the role of bicarbonate buffer in maintaining blood pH

The primary buffer system in plasma is bicarbonate, which is also the primary means of carbon dioxide transport in the body. Addition of hydrogen ions to bicarbonate ions produces carbonic acid, which buffers the added hydrogen ions so they do not contribute to changing the pH. Removal of hydrogen ions by bases is buffered by conversion of carbonic acid to bicarbonate, preventing any decrease in hydrogen ion concentrations and maintaining constant pH. Thus, the bicarbonate buffer system buffers both acid and base additions. Other buffer systems similarly soak up excess hydrogen ions or minimize decreases in hydrogen ions when bases are added. The most important secondary buffer in plasma is the protein albumin.

> Most important buffer in plasma is bicarbonate buffer system
> Addition of acids to bicarbonate produces carbonic acid which buffers pH
> Removal of hydrogen ions by base addition is buffered by bicarbonate

pH Balance

Explain the role of pulmonary ventilation in maintaining constant plasma pH

Metabolism produces acids. Extracellular acids are buffered by bicarbonate, producing carbonic acid. This carbonic acid is converted to carbon dioxide by *carbonic anhydrase* in red blood cells. Carbon dioxide is then removed by the lungs. When more acids are produced, carbon dioxide levels rise and the body responds by increasing ventilation to blow off the excess carbon dioxide and restore pH to its normal range. Thus, carbon dioxide is a means of getting rid of acids produced by metabolism. High carbon dioxide levels in the blood are associated with respiratory acidosis, the failure to maintain pH by ventilation.

> Carbon dioxide is in equilibrium with carbonic acid which removes acids
> Increased ventilation removes carbon dioxide, keeping pH from falling
> Kidneys produce bicarbonate that acts as a reserve of buffer

An optimal ratio of bicarbonate and carbonic acid is normally maintained moment-by-moment by adjusting the depth and rate of breathing. When plasma pH decreases, increased ventilation blows off more carbon dioxide, which is equivalent to getting rid of carbonic acid. Alternatively, hyperventilation decreases carbon dioxide levels below their optimal level, and pH increases. The kidneys contribute to long-term pH balance by secreting acids and producing bicarbonate. The bicarbonate produced by kidneys contributes to bicarbonate (alkaline) reserve that is used to buffer acids.

Arterial Blood Gases

Define and explain the normal limits of arterial blood gases

Only a narrow range of pH (7.35 to 7.45) is compatible with health. Plasma pH values outside this range are considered clinically relevant. *Acidosis* is a plasma pH < 7.35. *Alkalosis* is a plasma pH > 7.45. The pH of a buffer is given by the Henderson-Hasselbalch equation:\pH =

pKa + log ([A⁻] / [HA]), where [A⁻] is the concentration of salt (bicarbonate) and [HA] is the concentration of acid (carbonic acid). The pKa of bicarbonate buffer at body temperature is 6.1. To maintain pH 7.4 solely with bicarbonate requires a ratio of bicarbonate to carbonic acid of 20:1. For an optimal bicarbonate concentration of 24 mEq/L, normal pH is maintained when partial pressures of carbon dioxide are between 35 and 45 mm Hg.

Acid-base balance can be evaluated by measuring *arterial blood gases* (ABG). These measurements include blood pH and measure carbon dioxide concentrations in units of partial pressure ($PaCO_2$ in mm Hg). Bicarbonate concentrations are calculated in mEq/L by entering the measured values into the *Henderson-Hasselbalch equation*. Partial pressures of carbon dioxide can be converted to mEq/L by multiplying by 0.03. Thus, pH = 6.1 + log ([bicarbonate]/[0.03 x $PaCO_2$]).

The normal range of $PaCO_2$ is 35 to 45 mm Hg, which corresponds to a pH range of 7.35 to 7.45 when bicarbonate concentrations are 24 mEq/L. If pH and partial pressures of carbon dioxide are outside this normal range when bicarbonate levels are normal, a respiratory disturbance is suspected. A pH < 7.35 and high $PaCO_2$ (> 45 mm Hg) is *respiratory acidosis*. A pH > 7.45 and low $PaCO_2$ (< 35 mm Hg) is *respiratory alkalosis*.

> Plasma pH is given by Henderson-Hasselbach equation
> Optimal pH of 7.4 requires bicarbonate-carbonic acid (20:1 ratio)
> Outside this range is acidosis (pH < 7.35) or alkalosis (pH > 7.46)

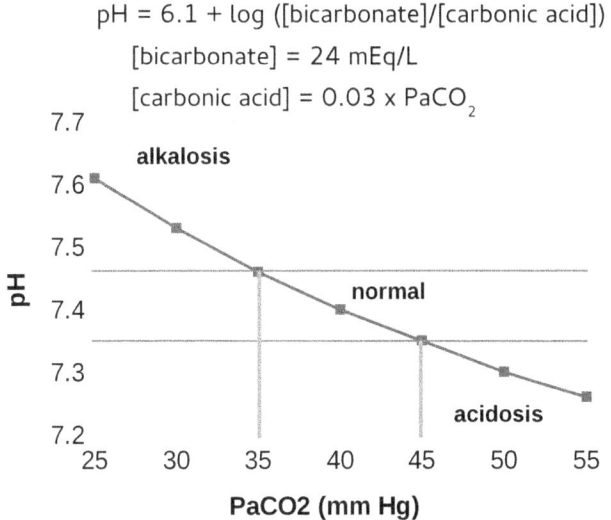

The kidneys compensate for chronic respiratory acidosis by producing bicarbonate. Thus, *compensated respiratory acidosis* is associated with a bicarbonate concentration above normal (> 26 mEq/L). Acute and chronic lung diseases, such as pneumonia or COPD, are the most common causes. Because the respiratory system is compromised, arterial oxygen levels are often low (PaO_2 < 80 mm Hg). Respiratory alkalosis is always the result of hyperventilation.

Review Questions

1. What is meant by partial pressure? What is Dalton's law?
2. What is Henry's law? Why are partial pressures used to describe gas concentrations in solution?
3. How do bronchi branch to form alveoli?
4. Describe the structure of the alveolus and the respiratory membrane.
5. What determines the direction of gas diffusion in the alveoli and tissues?
6. How is oxygen carried in the blood and exchanged with tissues?
7. In what forms is carbon dioxide transported in the blood?
8. How does ventilation maintain blood pH constant? What role do kidneys play in pH control?
9. What is the normal pH range? Why is bicarbonate-carbonic acid ratio important?

Renal Physiology

Your mind should be so imbued with physiological principles, that when disease is presented to you, you may at once regard it in its relations to the standard of health.

– Sir William Bowman

Renal Anatomy and Filtration

Metabolism, especially of proteins, produces wastes that enter the blood and must be removed from the body by the *kidneys*. The primary wastes that are removed are *urea, a* product of protein metabolism, *creatinine*, from muscle activity, and uric acid, from nucleotide metabolism. Failing kidneys do not fully remove these waste products, resulting in *uremia*, buildup of the most common waste product (urea) in the blood. In addition, the kidneys balance ion concentrations and pH and regulate blood volume.

The functional unit of the kidney is the *nephron*. Blood enters a tuft of capillaries (the *glomerulus*) in the outer layer of the kidney (*cortex*). Here it is filtered through a specialized epithelium into the proximal tubules. This filtrate contains nutrients, salts, soluble wastes, and water, but normally does not contain proteins. In the *proximal tubules*, most of the water and nutrients of the filtrate are reabsorbed. Additional water and salts are reabsorbed in the *loops of Henle* which dip into the inner layer of the kidney (*medulla*). Reabsorption in the first part of the nephron tubule (the proximal tubules and loop of Henle) is obligatory: it always occurs. In the *distal tubules*, reabsorption is regulated by hormones which contribute to acid-base balance and regulate blood volume by producing a more or less concentrated urine.

Filtrate from multiple nephrons empties into *collecting ducts* which pass into the center of the kidney (*renal pelvis*). Urine empties into ureters which carry it to the *urinary bladder* for temporary storage. When the bladder is full (and the time is right), urine is emptied from the bladder through the urethra.

Urinary System

Describe the functions of the urinary system

In the urinary system, the *kidneys* form urine, pass it to the *urinary bladder* for storage, and excrete it through the *urethra*. Kidneys filter blood that passes through them, then reabsorb what the body needs. Filtration removes most wastes. Reabsorption recovers filtered materials, including nutrients (glucose and amino acids), phosphate, calcium, magnesium and water. Kidneys also maintain long-term acid-base balance by removing acids and reabsorbing bicarbonate. They regulate volume and composition of body fluids by maintaining potassium and sodium balance and by secretion of the hormone renin.

> Kidneys filter blood and reabsorb what the body needs to form urine
> Urine is passed to bladder for storage and excretion
> Kidneys also regulate volume and composition of body fluids

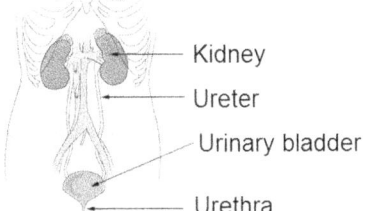

Kidney
Ureter
Urinary bladder
Urethra

Kidneys

Describe the structure of the kidneys and the pathway for urine

The kidneys are *retroperitoneal* (posterior to the peritoneal membrane of the abdominal cavity). They are covered with a *renal capsule* that helps maintain shape. Adipose tissue that cushions and supports them. *Renal fascia* anchors the kidneys to the posterior abdominal wall. *Renal cortex* is the outer layer which extends deeply as renal columns. The *renal medulla* is the inner layer of the kidney. The medulla is organized into *renal pyramids* that end in *renal papillae*. Together, renal pyramids and cortex form lobes. The *renal hilum* is the entry site for the ureters and blood vessels.

> Cortex is outer layer of kidney that extends into medulla as renal columns
> Renal pyramids: functional regions in medulla; apex is renal papilla
> Urine empties into minor calyx, major calyx, pelvis, ureter

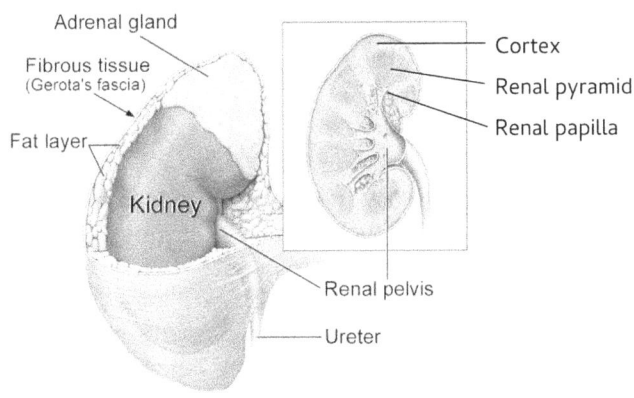

Adrenal gland
Fibrous tissue
(Gerota's fascia)
Fat layer
Kidney
Cortex
Renal pyramid
Renal papilla
Renal pelvis
Ureter

Blood Supply to Kidneys

Describe the supply of blood to kidneys and renal cortex

Blood enters the kidney through the *renal artery*, which branches into *segmental arteries*, then *interlobar arteries* that pass between the renal pyramids, *arcuate arteries* that arch over the pyramids between medulla and cortex, and *cortical radiate arteries* that radiate outward into the cortex.

> Renal arteries supply and renal veins drain blood from kidneys
> Interlobar pass between pyramids; arcuate along medulla-cortical boundary
> Cortical radiate arteries extend into cortex to supply nephrons

Nephrons are the functional units of the kidneys. Cortical radiate arteries send out *afferent arterioles* that enter the glomerulus, a tuft of capillaries that filter blood. After filtration, blood leaves the glomerulus through *efferent arterioles* that branch to form *peritubular capillaries* that surround the tubules of the nephrons. Straight tubular capillaries that dip deep into the medulla are *vasa recta*. Blood from tubular arteries collect into veins with the same names as the arteries: cortical radiate, arcuate, interlobar, segmental, and renal veins.

Nephrons are complex tubes that extend from the cortex into the medulla and back into the cortex. Nephrons that do not dip deeply into the medulla are *cortical nephrons*. Those that enter deeply into the medulla are *juxtamedulllary* nephrons, so named because their glomeruli are in the cortex at the border of the medulla. Tubules from multiple nephrons empty into *collecting ducts* which merge into *papillary ducts* that empty into renal papillae. The urine passes through minor and major calyces to empty into the renal pelvis.

> Afferent arterioles supply glomeruli; efferent arterioles leave them to
> become peritubular capillaries that dip deep into medulla
> Nephrons empty into collecting ducts then into renal papilla

Divisions of the Nephron

Describe the divisions of the nephron and their functions

Nephrons have five divisions: renal corpuscle, proximal convoluted tubules, loop of Henle, distal tubules, and collecting ducts. *Bowman's capsule* captures filtrate from the glomerulus which empties into the proximal convoluted tubule. Together with the glomerulus, the capsule forms the *renal corpuscle*. Most reabsorption of nutrients and water takes place in the *proximal convoluted tubules* (PCT), in the cortex.

Tubules then dip into the medulla and return to the cortex as the loop of Henle.

In the *loop of Henle,* additional water and salts are reabsorbed. When the nephron loops back up into the cortex, the tube passes next to the afferent arteriole of the glomerulus where the *distal convoluted tubule* (DCT) begins. In the distal tubules, reabsorption of water and salts is regulated by hormones.

The distal tubules of several nephrons empty into *collecting ducts* which passively reabsorb water to form a concentrated urine when the body needs to conserve water. Collecting ducts empty into papillary ducts that direct urine to the minor calyx which empties into a major calyx and the renal pelvis.

> Bowman's capsule collects filtrate from glomerulus and empties into PCT
> PCT reabsorbs nutrients and water; loop of Henle reabsorbs water
> Water reabsorption is under hormonal control in DCT and collecting ducts

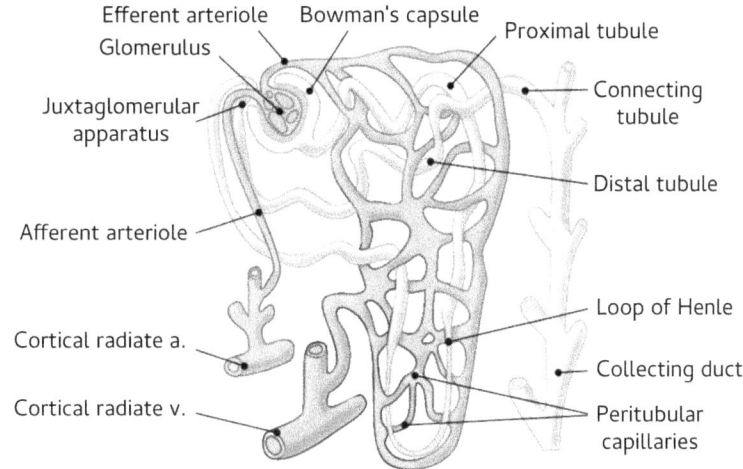

Urine Formation

Describe the process of urine formation

The process of urine formation may be divided into three components: 1) *filtration* of blood from glomerulus into Bowman's capsule; 2) *reabsorption* into peritubular capillaries of nutrients plus water and salts as needed to maintain homeostasis of body fluids; 3) *secretion* of certain poisons and acids into tubular fluids. *Excretion* refers to materials that remain in the final urine and are excreted from the body.

The excretion of any individual substance is determined by the amount filtered minus the amount reabsorbed plus the amount secreted: Excretion = filtered – reabsorbed + secreted. Substances needed by the body are reabsorbed; waste products are secreted. Filtration rate is maintained nearly constant. The filtrate is nearly identical in composition to plasma without proteins. Reabsorption of nutrients is nearly complete. Reabsorption of other substances is regulated to preserve body fluid balance.

> Blood is filtered from glomerulus into Bowman's capsule
> Nutrients and water are reabsorbed into peritubular capillaries
> Poisons and acids are secreted into tubules for excretion in urine

Glomerular Filtration

Describe the structure of the glomerular filtration membrane and how it filters blood

The *glomerulus* is a tuft of capillaries inside Bowman's capsule. It is covered with specialized cells (*podocytes*) that send out foot processes that partly cover the endothelium of the glomerular capillaries. Blood pressure forces fluids across the selectively permeable membrane (*glomerular filtration*) formed by the endothelium and podocytes. The resulting *glomerular filtrate* in Bowman's capsule is plasma without cells or proteins.

> Glomerulus is tuft of capillaries covered by podocytes
> Filtrate (no cells or proteins) enters Bowman's capsule
> and passes into proximal convoluted tubule

Glomerular filtrate directly enters the proximal convoluted tubule from the capsular space. Bowman's capsule is covered with a squamous epithelium. The entrance to the proximal tubule is distinguished by transition to a cuboidal epithelium. In a cross-section, proximal and distal tubules cannot be distinguished from each other. All renal corpuscles are in the cortex.

The *glomerular filtration membrane* is a specialized filtration barrier composed of endothelial cells, basal lamina (a shared basement membrane), and podocytes. The endothelial cells of the glomerulus are *fenestrated capillaries* (with pores) that prevent filtration of cells.

> Glomerular capillaries are fenestrated: endothelial cells are connected by
> tight junctions and have small holes in cells that act as filtration barrier

Filtration slits are thin membranes formed by the foot processes of podocytes. Blood plasma is filtered through endothelial fenestrations, which keep out cells. Then filtrate passes through the basal lamina and filtration slits, which keep proteins out of the filtrate. Proteins do not cross the basal lamina because they are large and the filtration membrane has many negatively charged glycoproteins that repel the negatively charged plasma proteins. Glomerular filtrate has the composition of plasma, but without its proteins.

> Glomerular filtration membrane consists of fenestrated glomerular capillary endothelium and shared basal lamina secreted by endothelial cells and foot processes of podocytes

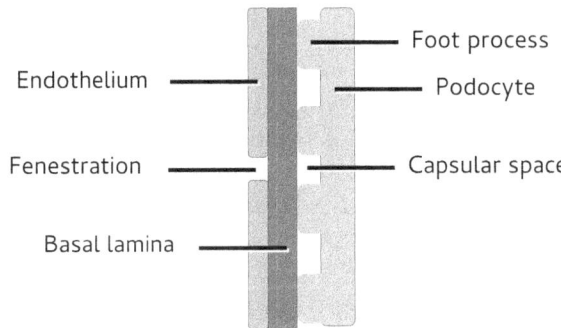

Net Filtration Pressure

Define the pressures that influence glomerular filtration

Blood pressure forces fluids from glomerular capillaries into Bowman's capsule. This is glomerular filtration. About 20% of plasma fluids are filtered into Bowman's capsule (*filtration fraction*). In the renal corpuscle, blood pressure is *glomerular hydrostatic pressure* (GHP) and the osmotic pressure of the blood is *blood colloid osmotic pressure* (BCOP). As in other capillaries, the force pushing fluids out (GHP) is partly balanced by osmotic pressure of proteins in the capillary that pulls fluids back in (BCOP).

In most tissues, little resistance to the outward force is encountered. In the renal corpuscle, back-pressure builds as fluids enter the small space of Bowman's capsule. This is the *capsular hydrostatic pressure* (CHP). A typical value for blood pressure in the capillaries (GHP) is close to 45 mm Hg. The opposing osmotic pressure in capillaries (BCOP), due to its much higher protein concentration, is about 19 mm Hg at the beginning of the glomerular capillary. Loss of fluids across the glomerulus increases osmotic pressure to about 35 mm Hg near the end of the glomerular capillary. Capsular hydrostatic pressure is about 10 mm Hg.

Glomerular hydrostatic pressure: blood pressure forces fluids out
Capillary osmotic pressure: pulls fluids back in due to blood proteins
Capsular hydrostatic pressure builds up in filtration space to oppose flow

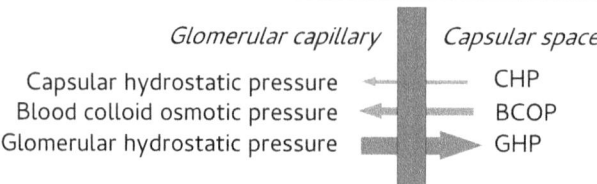

Glomerular capillary		*Capsular space*
Capsular hydrostatic pressure	←	CHP
Blood colloid osmotic pressure	←	BCOP
Glomerular hydrostatic pressure	→	GHP

The balance of these Starling forces is net filtration pressure: NFP = GHP – BCOP – CHP. NFP is typically 16 mm Hg where the afferent arteriole enters the glomerulus and falls close to zero near the exit of the efferent arteriole. Thus, blood in tubular capillaries has high on-cotic pressure which promotes fluid reabsorption in the nephron.

Glomerular Filtration Rate

Describe the factors that influence GFR

The volume filtered per unit time is the *glomerular filtration rate* (GFR). Typically, GFR is close to 125 mL/min. This means that the blood of the entire body is filtered through the kidneys some 60 times a day! GFR is determined by the permeability and total area of the filtra-tion membrane (filtration coefficient) and net filtration pressure (NFP). In the absence of disease that disrupts the filtration barrier or reduces the number of functional nephrons, the filtration coefficient is constant. Thus, GFR is normally determined by net filtration pressure.

Vasoconstriction of afferent arteriole reduces blood flow through glomerulus and decreases blood pressure to reduce GFR
Efferent arteriole vasoconstriction increases local BP to increase GFR

Vasoconstriction of   Vasoconstriction of
afferent arteriole   efferent arteriole

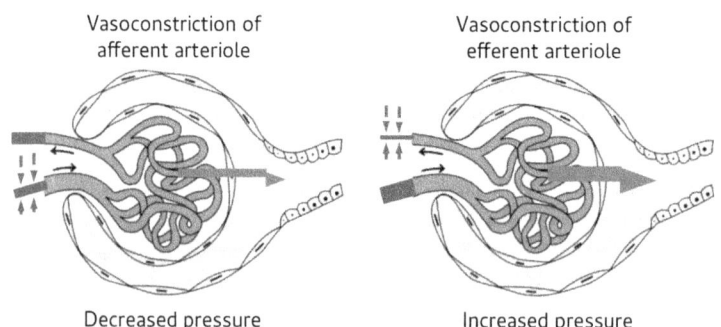

Decreased pressure   Increased pressure

If large amounts of protein are lost in the urine, due to damage to the filtration membrane (as in nephrotic syndrome), blood colloid osmotic pressure decreases and net filtration pressure and GFR increase. But in the normal kidney, blood pressure is the primary determinant of GFR. What matters for filtration is the local pressure in the glomerular membranes. If systemic BP changes, the kidneys compensate by altering blood flow through the glomeruli to maintain constant GFR.

When systemic BP increases, blood flow to the glomerulus is decreased by vasoconstriction of the afferent arteriole. This reduces the local blood pressure in the glomerulus, maintaining a relatively constant net filtration pressure. When systemic BP decreases, blood flow out of the glomerulus is restricted by vasoconstriction of the efferent arteriole. This increases local blood pressure in the glomerulus, maintaining constant net filtration pressure.

> GFR remains nearly constant over a wide range of systemic blood pressures
> As BP increases, vasoconstriction of afferent arterioles reduces
> glomerular capillary pressure, NFP, and restores GFR to normal range

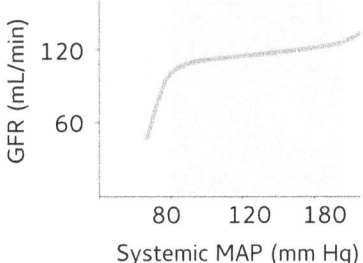

Tubuloglomerular Feedback

Describe how GFR is autoregulated to remain relatively constant

The primary mechanism that maintains constant GFR is a (local) paracrine response. The *macula densa* is a set of closely packed cells lining the wall of the distal tubule next to the arterioles of the renal corpuscle. This cluster of cells is named for their prominent nuclei, which makes them appear darker in stained sections. These cells monitor filtrate that has passed through the proximal convoluted tubules and the loop of Henle. This filtrate should have a constant composition since GFR is autoregulated to remain nearly constant and reabsorption in the proximal tubules and loop of Henle is steady and unchanging.

Macula densa cells respond to changes in NaCl concentration in the tubular fluids entering the distal tubules. This response feeds back to the arterioles surrounding the glomerulus, so is called *tubuloglomerular feedback*. For example, a slight increase in GFR causes a higher rate of fluid flow through tubules. This rapid flow reduces the amount of NaCl reabsorbed in the proximal convoluted tubule and ascending limb of the loop of Henle. Macula densa cells sense the increase in salt concentration in the tubular fluid and release adenosine in response.

Adenosine binds to its receptors on smooth muscle cells of glomerular afferent arterioles causing vasoconstriction. Vasoconstriction reduces blood flow into the glomerulus to normal levels, in spite of the increased systemic blood pressure. This decrease net filtration pressure to maintain a constant GFR.

> Increased GFR decreases NaCl concentration of fluid in DCT
> leading to adenosine release from macula densa
> Adenosine causes vasoconstriction of afferent arterioles

At extremes of low and high blood pressure, GFR is no longer autoregulated. At very low blood pressures (due to excessive fluid loss), GFR decreases to prevent cardiovascular collapse. This is the result of sympathetic innervation that constricts the afferent arteriole, reducing blood flow and decreasing GFR. At very high blood pressures, GFR increases slightly in spite of compensatory vasoconstriction of afferent arterioles.

Renal Blood Flow

Explain the importance of sufficient renal blood flow

Kidneys receive some 25% of total cardiac output. Sufficient renal blood flow (RBF) is critical to maintaining homeostasis by removing wastes and regulating the composition of body fluids. Blood flow corrected for hematocrit gives plasma flow. With a hematocrit of 40, renal plasma flow (RPF) is 60% of RBF. The fraction of plasma that is filtered into Bowman's capsule is given by the ratio of renal plasma flow and glomerular filtration rate: *filtration fraction* = RPF / GFR. Typical values for filtration fraction are close to 0.2, meaning 20% of plasma entering the glomerulus ends up in glomerular filtrate.

Renal artery stenosis (narrowing of the artery, often due to atherosclerosis) reduces blood flow to the kidneys. The kidneys respond to the reduced blood flow by releasing hormones that increase water retention leading to increased systemic blood pressure (renovascular hypertension). Obstruction of blood flow may also occur in the arterioles, causing local ischemia that damages tubular tissue. Systemic hypertension and diabetes mellitus are the primary risk factors for chronic kidney disease. Hypertension can directly damage kidney tissues, while diabetes mellitus causes glycosylation of the glomerular filtration membrane, compromising its function. Functioning nephrons attempt to compensate for damaged nephrons by increasing glomerular capillary pressure that damages the glomerulus, leading to progressive renal failure.

> Adequate blood flow to kidney is required to maintain kidney function
> Severely reduced blood flow to kidney causes ischemic damage (necrosis)

Renal Clearance

Define clearance and how it is used to estimate GFR

Clearance is the volume of plasma from which a substance is completely removed in one minute. Renal clearance measures how fast a drug or other plasma component is removed from the blood. Glucose clearance is zero: all of the glucose in renal filtrate is reabsorbed, so none is removed from plasma by the kidneys. Albumin clearance is also zero, because it is not filtered and does not enter the tubules.

If a substance is only filtered (not reabsorbed or secreted) its clearance equals GFR. That is, all of the substance is filtered out as it passes through the glomerulus. *Creatinine clearance* is nearly ideal for measuring GFR. Creatinine is filtered in the glomerulus, not reabsorbed, and only a tiny amount is secreted. Creatinine is produced by skeletal muscles at a relatively constant rate that depends only on muscle mass. This maintains a relatively constant level of creatinine in the blood. The amount of creatinine excreted is determined almost entirely by its rate of filtration. That is, creatinine clearance nearly equals GFR.

> Clearance measures volume of plasma cleared of a substance
> Glucose is completely filtered and reabsorbed; clearance is zero

To calculate GFR, blood concentrations of creatinine (C_B) are compared with those in the urine (C_U). Knowing the rate of urine production (Q in mL/min) gives GFR = C_U / C_B × Q. A typical urine volume is 1 mL/min (1440 mL/day). GFR is corrected for body surface area, using a standard surface area of 1.73 m^2. The normal range of GFR is 100-130 mL/min/1.73m^2 for adult men and 90-120 for women under 40. Often, only serum creatinine levels are measured and GFR is estimated by one of a number of standard formulas.

> Creatinine is produced at a constant rate by skeletal muscles
> Creatinine is completely filtered and not reabsorbed
> Creatinine clearance estimates GFR (because of some secretion)

Urinary Bladder

Describe the structure of the urinary bladder

Micturition is urination. Distal tubules of multiple nephrons pass fil-trate to collecting ducts which merge to form papillary ducts that empty into the renal pelvis to enter the ureters. The ureters carry urine (by peristalsis) to the urinary bladder for storage. The ureters consist of transitional epithelium with circular and longitudinal muscle layers. They empty into the urinary bladder, which has a similar tissue struc-ture.

> Urinary bladder consists of transitional epithelium folded into rugae
> Detrussor molecule contracts to push urine through urethra
> Internal and external urethral sphincters control micturition

The maximum volume of the urinary bladder is about 700-800 mL. Rugae (folds) in the mucosa allow the bladder to expand. The detrussor muscle contracts to expel urine which passes smoothly through the tri-angular-shaped trigone at the base of the bladder into the urethra. Micturition is controlled by internal and external urethral sphincters.

Micturition

Describe the micturition reflex

When the bladder is partly full (about 250 mL), stretch receptors in its wall begin to respond. A parasympathetic reflex causes detrussor mus-cle contraction in the bladder wall. Urine is pushed toward the urethra, forcing the internal urethral sphincter to open. Urine is channeled into ureters through the smooth trigone. The external urethral sphincter is under voluntary control and must relax to allow passage of urine. The urethra is short in females and longer in males.

> Stretch receptors respond with parasympathetic reflex when bladder full
> Voluntary control of external urethral sphincter controls urination

Review Questions

1. What are the main components of the urinary system and their functions?
2. How are the kidneys protected in the abdominal cavity?
3. What are the layers of the kidneys and its primary internal structures?
4. What are the divisions of the arteries and veins that supply and drain the kidneys?

5. What are the two types of nephrons?

6. How is blood supplied to nephrons and their tubules?

7. What is the path from nephron tubules to renal pelvis?

8. What are the five divisions of the nephron and their functions?

9. What are the three processes that occur in nephrons to form urine?

10. What is a glomerulus? Describe the structure of the glomerular filtration membrane.

11. What forces drive fluids across the filtration membrane of the renal corpuscle?

12. What factors influence glomerular filtration rate (GFR)?

13. Why is sufficient renal blood flow required for effective kidney function?

14. What is filtration fraction and its typical value?

15. What are the mechanisms that regulate glomerular filtration rate?

16. What is the effect of systemic blood pressure on GFR?

17. What is clearance and how is creatinine clearance used to estimate GFR?

18. What is the function of the rugae, trigone, and detrussor muscle of the urinary bladder?

19. How do the internal and external urethral sphincters differ?

20. What is the micturition reflex?

21. How is urination under voluntary control?

Renal Reabsorption Mechanisms

Blood is filtered from capillaries in the glomerulus into Bowman's capsule. This filtrate does not contain cells or proteins, but does have nutrients that the body needs and does not want to lose in the urine. One of the most important functions of the nephrons is to recover nutrients and water from the filtrate by reabsorption into the capillaries. Some wastes are not filtered but are secreted from the peritubular capillaries into the nephrons. *Reabsorption* of water, salts, and ions is regulated by hormones (mostly *ADH* and *aldosterone*) in the distal tubules and collecting ducts. What remains in the tubules is excreted as urine.

In this lecture, we examine the mechanisms that account for reabsorption in different parts of the nephron and collecting ducts. These include *transporters* and *channels* for ions and small molecules. Sodium ions are always pumped out of the PCT and ascending limb of the loop of Henle. Salts are pumped out of the DCT when aldosterone is present. Water reabsorption always follows Na reabsorption in the PCT. In the descending limb of the loop of Henle, water reabsorption is driven by the salt gradient established by Na pumps in the ascending limb. The distal tubules and collecting ducts are only permeable to water when ADH is present. Thus, ADH causes urine to be more concentrated because it allows water reabsorption in the distal parts of the nephron and collecting ducts.

Scheme of Reabsorption

Describe where nutrients, salts, and water are reabsorbed in the nephron

Each segment of the nephron has a different set of transporters that determine what substances are reabsorbed. In the first part of the nephron (proximal tubules and loop of Henle), reabsorption is *obligatory*: it always occurs and does not require hormones. Most of the reabsorption occurs in the proximal tubules. Some additional water is reabsorbed in the descending limb of the loop of Henle. NaCl reabsorption in the ascending limb establishes a salt gradient wherein osmolarity is very high deep in the medulla.

Reabsorption of water, salts, and ions in the late nephron (distal tubules and collecting ducts) respond to hormones. ADH (antidiuretic hormone) is required for water permeability. Aldosterone acts on distal tubules to cause salt reabsorption that drives water reabsorption by osmosis. Urea is reabsorbed in the deep medulla, contributing to its high osmolarity.

> PCT reabsorbs most water, salts, and soluble nutrients
> Loop of Henle reabsorbs water (descending limb) and NaCl (ascending limb)
> Distal tubules and collecting ducts reabsorb water when ADH is present

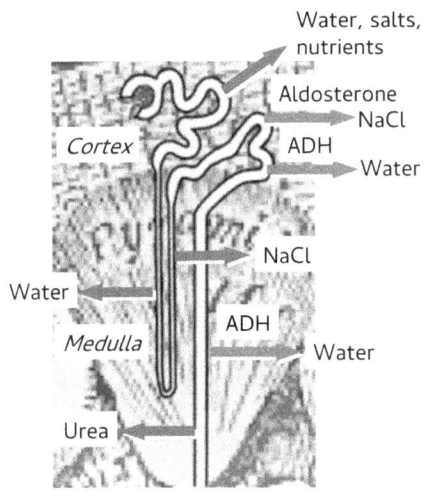

Secondary Active Transport

Describe mechanisms that drive nutrient and water reabsorption in nephrons

Both nutrient and water reabsorption in the nephron tubules is driven by active transport of Na ions from cells lining the nephrons into the interstitial space. This is due to an active *Na-K ATPase* on the basolateral surface of the cells (toward the interstitial fluids). Pumping of Na ions increases Na concentrations in interstitial fluids and reduces Na ion concentrations within the cells. The Na pump essentially "pulls" Na from the tubular lumen across the lining cells into the surrounding fluids, where it is free to enter the peritubular capillaries to be carried away from the nephrons.

The concentration gradient thus established can drive water reabsorption from the tubules into the interstitial fluids by *osmosis*. It can also be coupled to reabsorption of glucose and amino acids, which are *co-transported* with Na ions through the apical membrane facing the tubular lumen. This coupling of active transport of Na ions by means of the Na pump, with co-transport of nutrients into the cells is *secondary active transport*. Na ions pass down the concentration gradient established by the basolateral Na pump. Reabsorption is not directly coupled to energy-dependent Na transport but depends on the concentration gradient established by active transport.

Na pump in basolateral membrane establishes Na gradient
Nutrients cotransported with Na ions and water by osmosis
from tubular fluids across cells to interstitial fluid

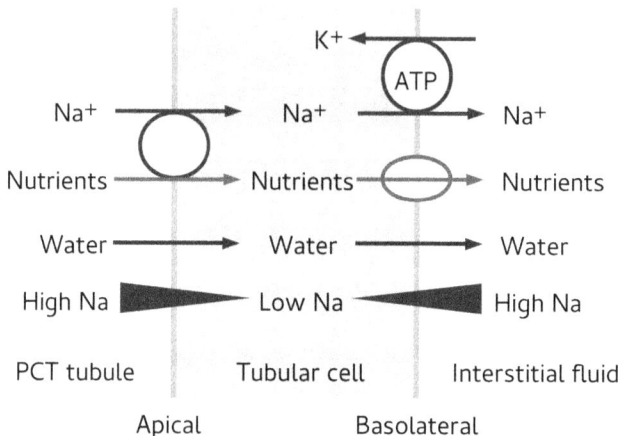

Nutrient Reabsorption

Describe the mechanisms by which nutrients are is reabsorbed in proximal tubules

Most of the water and essentially all of the nutrients (largely amino acids and glucose) in Bowman's capsule filtrate are reabsorbed in the *proximal convoluted tubules*. Reabsorption of nutrients in the PCT uses mechanisms similar to those that account absorption of nutrients in the small intestine. Cells lining the PCT are cuboidal epithelium with microvilli that provides a large surface area for reabsorption. The reabsorption mechanisms are highly efficient: nearly all the glucose and amino acids in the filtrate are reabsorbed.

Amino acid transport is carried out by a wide variety of apical (luminal) and basolateral transporters coupled to Na and H ion transport. Most transporter systems can carry several different amino acids and individual amino acids are can be carried by more than one transporter. Small proteins in the filtrate enter proximal tubular cells by endocytosis, are degraded in lysosomes to amino acids, and transported across the basolateral membrane by amino acid transporters. Lactate and intermediates in the citric acid cycle, such as citrate and α-ketoglutarate, are also reabsorbed by secondary active transport in the PCT.

Glucose transport is driven by Na gradient established by Na pump
Glucose enters proximal tubule cells by Na-linked glucose transporters
and passively transported into interstitial fluids by glucose transporter 2

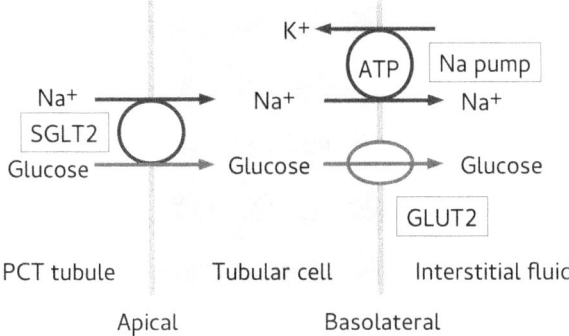

The primary apical transporter for glucose is *SGLT2* (sodium-glucose linked transporter type 2), which reabsorbs 90% of the glucose in the initial segment of the proximal convoluted tubule. The remainder is reabsorbed via a closely-related SGLT1 transporter. Glucose that accumulates in the cells then passes into the interstitial fluids by facilitated transport through specific transporters for glucose (GLUT2) on the basolateral membrane.

Glucose transport is limited by capacity of SGLT transporters
When filtrate concentrations are very high glucose spills into urine

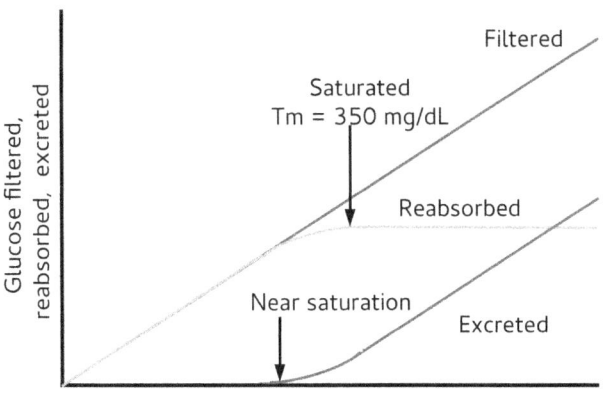

Glucose reabsorption is limited by the capacity of the apical SGLT transporters. If glucose levels in the blood are too high (hyperglycemia), as in uncontrolled diabetes mellitus, not all glucose in the filtrate is reabsorbed and what remains is excreted to produce a sweet urine (*glucosuria*).

The transporter is near saturation at about 180 mg/dL of glucose (the renal threshold for glucose). At this level, small amounts of glucose are spilled into the urine. The transporter maximum for glucose (Tm) is about 350 mg/dL. Above that level all excess glucose spills into the urine. Pharmacological inhibitors of SGLT2 (*gliflozins*) increase glucose excretion, thereby reducing blood glucose levels in type 2 diabetics.

Water Reabsorption in PCT

Describe the processes that cause reabsorption of water in PCT

Water is reabsorbed throughout the tubular system by osmosis due to a salt gradients established by Na pumps. Most water reabsorption occurs in the proximal tubules. Additional water is reabsorbed in the loop of Henle. Water reabsorption in the distal convoluted tubule is hormonally regulated in response to changes in blood volume, osmolarity, and pressure.

> Reabsorption of water is driven by osmosis
> Always occurs in PCT and descending limb of loop of Henle
> Regulated by hormones in distal tubules and collecting ducts

The driving forces for water reabsorption are osmotic. In the proximal convoluted tubules, Na is pumped from the tubular epithelial cells into the interstitial fluids causing an osmotic gradient for water reabsorption. In addition, the oncotic pressure of the peritubular fluids is increased because of the fluid loss by glomerular filtration. In the PCT, water reabsorption is isosmotic: it comes to equilibrium with the surrounding fluids.

Normally two-thirds of the Na ions in the tubule are reabsorbed, along with the same fraction of water, in the PCT. Lesser amounts are reabsorbed in the loop of Henle. The remainder of water reabsorption is under hormonal control in the distal tubules.

Loop of Henle

Explain how the loop of Henle established a medullary salt gradient and reabsorbs water

The *loop of Henle* lies between the proximal and distal tubules. Its primary function is to establish a *medullary salt gradient* that drives water reabsorption by osmosis. It is most distinct in juxtamedullary nephrons which have a loop of Henle that deeply penetrates into the medulla.

Only the *descending limb* of the loop of Henle is permeable to water. Salts are pumped out of the *ascending limb* to increase osmotic concentration in the renal medulla. As tubular fluids pass down into the medulla within the descending limb, water is reabsorbed by osmosis. By the time it reaches the depths of the medulla, where the loop is located in juxtamedullary nephrons, the osmolarity is on the order of 900 mOsM. On the way back to the cortex, salts in the tubular fluids are reabsorbed to bring the osmolarity back to about 100 mOsM.

> NaCl is pumped out of ascending limb of loop of Henle
> to establish osmotic gradient in medulla
> Water leaves descending limb as it passes deeply into medulla

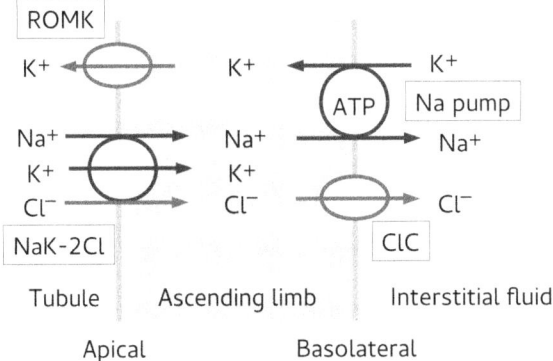

In the ascending limb, both Na and Cl ions are reabsorbed. A basolateral Na-K ATPase pumps Na into the peritubular space. *ROMK* (renal outer medullary potassium) is a K ion leak channel in the luminal membrane. Na, K, and Cl ions in the tubular fluids enter the cell through the NaK-2Cl transporter (*NKCC2*) which is inhibited by loop diuretics. Na ions pass down their concentration gradient because of the Na pump and the high Na ion concentration in the ascending limb. K and Cl ions are co-transported with Na ions and Cl ions leave the cell through a basolateral chloride channel (ClC).

> Na is pumped out of cells in the ascending limb of loop of Henle
> driving reabsorption of Na and Cl ions (K ions equilibrate)
> Cl ions are reabsorbed through basolateral chloride channels

Countercurrent Exchange

Explain how vasa recta help maintain medullary osmotic gradient

Vasa recta are straight capillaries that penetrate deeply into the medulla along the length of the loop of Henle in juxtamedullary nephrons. Blood flow in these capillaries is in a direction opposite to the flow of tubular fluids in the loop of Henle. Blood flow is slow in the vasa recta. As they pass into the deep medulla along the ascending limb of the loop of Henle, their plasma comes to osmotic equilibrium with the surrounding interstitial fluids.

Vasa recta flow in opposite direction to tubules of loop of Henle Opposite flows maintain gradient by countercurrent exchange Concurrent flow would tend to dissipate the medullary gradient

Since the direction of the two flows is opposite, this equilibration is *countercurrent exchange*. It maintains the osmotic gradient established by the Na pump of the ascending limb. If the flow was in the same direction, the lower osmolarity of plasma would tend to dissipate the medullary gradient (concurrent exchange).

Distal Tubules

Describe the mechanisms of sodium reabsorption in the distal tubules

The distal tubules are divided into several regions which have different sets of transporters and channels and vary in their responsiveness to the hormone aldosterone. The *distal convoluted tubule* (DCT) is that part of the nephron immediately downstream of the macula densa. Distal tubules are divided into two functionally distinct subsegments. The early DCT (DCT1) does not respond to aldosterone; the late DCT (DCT2), connecting tubule, and cortical collecting ducts respond to aldosterone.

In all of these distal regions, a Na pump (Na-K ATPase) is located on the basolateral membrane and a thiazide-sensitive NaCl cotransporter (*NCC*) is on the luminal membrane. Na ions are pumped out of the cells into the interstitial fluids, establishing a Na gradient that drives Na and Cl ion uptake through the luminal NCC cotransporter. Cl enters the interstitial space through Cl channels (ClC) and a K-Cl cotransporter (KCC4) in the basolateral membrane. K ion balance is maintained by K ion leak channels ROMK and BK (big/maxi K channel).

In distal tubules downstream, an additional luminal channel for Na ions (*ENaC*), sensitive to amiloride, is present. Here, Cl also passes from tubule to interstitial fluids between cells (paracellular transport). K ions are secreted into the lumen in response to the electrical imbalance caused by Na ion entry into epithelial cells of the distal tubules. Thus, K efflux depends on an active ENaC. Thiazide diuretics increase Na ion concentrations entering the late distal tubule (DCT2). This increases ENaC activity, which is coupled to K secretion, so thiazides

cause K loss in urine. By contrast, inhibition of ENaC by amiloride reduces Na reabsorption, increasing urine volume while also inhibiting
the coupled K secretion. Thus, amiloride is a K-sparing diuretic.

Distal tubule Na pump drives Na ions across tubule epithelial cells
NaCl enters cells from lumen through thiazide-sensitive NCC channel
K ions are secreted through ROMK and BK channels

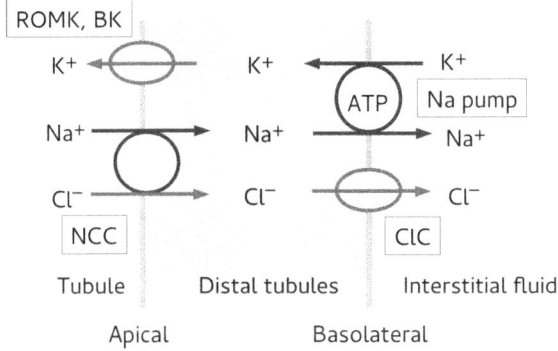

Aldosterone

Explain how aldosterone influences water reabsorption

Aldosterone is a steroid hormone (*mineralocorticoid*) produced by the
adrenal cortex in response to decreased blood pressure. It acts directly
and indirectly to increase activity of ENaC channels and the NCC symporter of the principal cells of the late distal tubule. This upregulation
increases Na reabsorption. Since K secretion is linked to ENaC activity, aldosterone also stimulates K secretion.

This aldosterone-stimulated Na reabsorption establishes a Na concentration gradient in the cortex that has the potential to drive water
reabsorption in the late distal tubules and cortical regions of the collecting ducts. However, the distal tubules and collecting ducts are
permeable to water only in the presence of antidiuretic hormone
(ADH). When ADH is secreted, the Na gradient established in the cortex drives water reabsorption by osmosis. Water reabsorption reduces
urine volume which increases blood volume and restores blood pressure to normal.

In late distal tubules ENaC channel contributes to Na ion reabsorption
and corresponding K ion secretion through leak channels
Aldosterone activates NCC symporter and ENaC channels

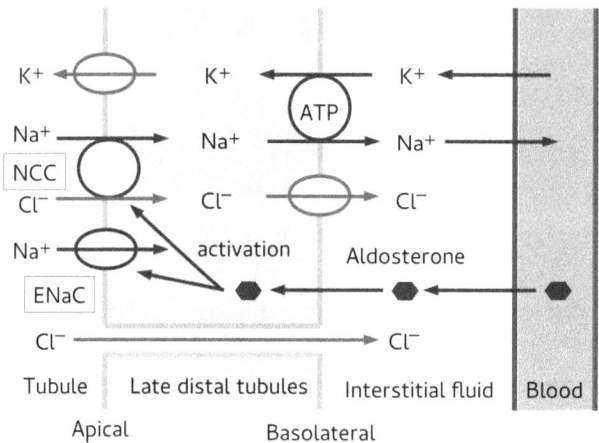

Antidiuretic Hormone

Explain the effects of antidiuretic hormones on water reabsorption

Water reabsorption in late distal tubules and collecting ducts
depends on aquaporins (water channels) inserted when ADH is present

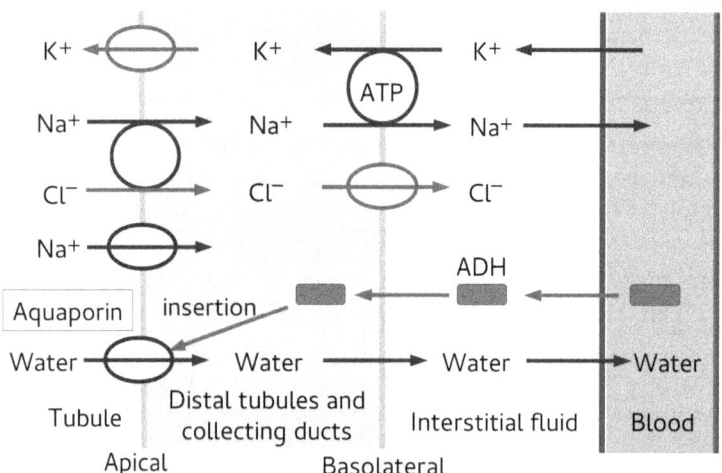

Water reabsorption is obligatory (always occurs) in the PCT and descending limb of the loop of Henle. In the distal tubules and collecting ducts, water reabsorption is regulated: it depends on hormones. Aldosterone increases Na reabsorption which increases interstitial Na concentrations in the cortex. ADH (antidiuretic *hormone* = vasopressin) causes insertion of aquaporin channels into distal tubules and collecting ducts, making them permeable to water.

When *aquaporin* channels are abundant, water is reabsorbed by osmosis as the collecting tubule enters the medulla. Here, osmolarity is high due to the NaCl pumps of the ascending limb of the loop of Henle. Water reabsorption occurs so long as the medullary osmotic gradient is maintained. Loop diuretics reduce this gradient, making them highly effective in increasing urine volume and reducing blood volume.

Urea Excretion

Describe the mechanisms that determine urea excretion

Ammonia is a by-product of amino acid metabolism. It is highly toxic, so ammonia is converted in the liver to the safer, water-soluble waste product *urea*. Urea is produced by combining two ammonia molecules with a molecule of carbon dioxide in the urea cycle. Urea enters the blood and is excreted by the kidneys. About 40% of the urea filtered is normally found in the final urine. In the PCT, urea is reabsorbed along with sodium and water. As water is reabsorbed, tubular urea concentrations increase and urea is reabsorbed by diffusion across the tubular epithelial membranes.

Urea is reabsorbed in PCT due to increasing tubular concentration
Urea is secreted into descending limb of loop of Henle
Urea concentration increases in medulla as urine is concentrated

As tubular fluids enter the descending limb of the loop of Henle, urea passes down its concentration gradient from interstitial fluids into tubular fluids. On the way back up the ascending limb, urea is retained, because the ascending limb is impermeable. Unlike salts that are reabsorbed in the ascending limb, which reduces their concentration in tubular fluids, the urea concentration remains high as the fluids enter the distal tubules.

Only the deep medullary part of the collecting ducts (the papillary region) is permeable to urea. In the presence of ADH, water is reabsorbed as the the tubular fluids pass down the collecting ducts. This further increases the tubular urea concentration. By the time it reaches the papillary region of the collecting ducts, tubular urea concentration is high. For this reason, urea is reabsorbed deep in the medulla where it contributes to the medullary osmotic gradient. When ADH levels are low, tubular urea concentrations do not increase significantly, and little urea is reabsorbed.

Secretion of Weak Acids and Bases

Describe the excretion of organic acids and bases

Small organic molecules are filtered through the glomerulus. Many are also secreted in the proximal convoluted tubules. Organic anions enter tubular cells through *organic anion transporters* (OAT) in the basolateral membrane in exchange for a dicarboxylic acid, such as glutarate. The dicarboxylic acid re-enters the cell through a transporter coupled to Na ion uptake (sodium-dicarboxylate cotransporter NaDC3). The anions enter tubules through ATP-dependent efflux pumps (*ABC transporters*) with broad substrate specificity. *Penicillin* is both filtered and secreted into the PCT by the organic anion transporter. *Probenicid* inhibits the transporter and can be given with penicillin to reduce its excretion. Organic cations, such as *morphine*, are transported by organic cation transporters (OCT).

> Organic acids and bases enter proximal tubular cells through OAT and OCT
> Both are secreted by ABC transporters (such as P-glycoproteins)
> Acidic urine traps bases; basic urine traps acids

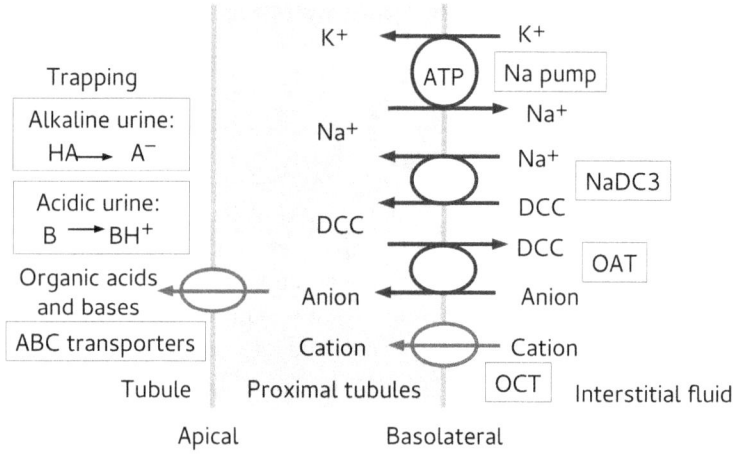

When the urine is acidic, transported anions are protonated and uncharged. In this form, they are reabsorbed because they can diffuse across cells of the tubules, down their concentration gradient. At elevated urine pH, weak bases are unprotonated and uncharged. Thus, weak acids (such as *salicylate*) are reabsorbed when urine pH is low and weak organic bases (such as quinine or morphine) are reabsorbed when urine pH is alkaline. So, excretion of salicylate is increased in alkaline urine. For morphine, a weak base, excretion is increased with acidic urine. Thus, clearance of aspirin or morphine can be increased in overdoses by changing urine pH (sodium bicarbonate increases pH; ammonium chloride decreases pH).

Review Questions

1. How does the basolateral sodium pump drive reabsorption of water and nutrients in renal tubules?
2. What transporters are involved in renal glucose reabsorption and where are they located?
3. Why does glucose spill into the urine of uncontrolled diabetics?
4. What mechanism accounts for water reabsorption?
5. In which parts of the kidney is water reabsorption obligatory and where is it regulated?
6. How does the ascending limb of the loop of Henle establish the medullary salt gradient?
7. What drives water reabsorption in the descending limb of the loop of Henle?
8. What is countercurrent exchange and how does it maintain the medullary osmotic gradient?
9. How is sodium reabsorbed in the distal tubules?
10. How does aldosterone increase sodium reabsorption in the distal tubules?
11. What additional hormone is required for aldosterone to increase water reabsorption in the distal tubules?
12. Why is ADH required for water reabsorption in the distal tubules and collecting ducts?
13. What drives urea reabsorption in the PCT?
14. Under what conditions is urea reabsorbed in the papillary ducts?
15. What is the effect of papillary duct urea reabsorption on the medullary osmotic gradient?
16. What drives Ca reabsorption in the distal tubules?
17. How are organic acids and bases secreted in the PCT?
18. How does urine pH influence excretion of organic acids and bases?

Body Fluids

A major function of the kidneys is to rid the body of soluble wastes in blood. It also plays a central role in maintaining blood pressure and balancing ion concentrations, including blood pH. Nutrients and water are largely reabsorbed in the proximal tubules and loop of Henle. Salt reabsorption (especially of Na ions) largely drives this reabsorption.

Adjustments to the urine are made in the distal tubules and collecting ducts. These adjustments are a response to blood pressure and volume, which determines whether the urine will be dilute (thereby reducing blood pressure) or concentrated (which increases blood pressure by preserving water). The distal tubules also ensure that the blood does not lose too much calcium by reabsorbing it from the filtrate. Long-term regulation of pH is accomplished primarily by secreting acids and reabsorbing bicarbonate to prevent acidosis.

In this lecture, we begin with a review of *body fluids* and the compartments into which they distribute. Here, what is most important is regulation of *blood volume* and *osmolarity* (primarily through regulating Na ion concentrations). Then, we look into control mechanisms that regulate secretion of *ADH* and *aldosterone* which largely influence urine concentration. Opposing these is *ANP*, which is produced by the heart when atrial fill volume increases as a result of increased venous return pressure. It has effects opposite to those of aldosterone.

We look at the relationships between Na reabsorption in distal tubules and K secretion and its effects on *blood potassium* levels. *Ca levels* are primarily regulated by PTH (*parathyroid hormone*). PTH mobilizes calcium from bones, increases renal calcium reabsorption, and stimulates dietary Ca absorption. The lecture concludes with a discussion of *acidosis* and *alkalosis* and *compensatory mechanisms* the body brings into play to maintain pH in the face of failing renal or respiratory function.

Body Fluid Compartments

Describe the major fluid compartments of the body, state their volumes, and describe their composition

About 60% of body weight is water. *Intracellular fluids* (ICF) make up about 2/3 of body fluids. *Extracellular fluids* (ECF) are either *intersti-*

tial fluids (in tissues) or *intravascular fluids* (in blood vessels). Interstitial fluids account for about three-fourths of the ECF.

> Most body water is in intracellular fluids (inside cells, ICF)
> Extracellular fluids (ECF) are largely interstitial (between cells) or plasma
> ICF contains K and phosphate ions; ECF has Na and Cl ions

Electrolytes in intracellular fluids are dominated by K and phosphate ions. Extracellular fluids contain predominantly Na and Cl ions. Plasma and interstitial fluids are in *free exchange*, except for proteins which are more concentrated in plasma than interstitial fluids. High protein concentrations within cells are balanced by pumping of Na ions out of cells to maintain *osmotic balance* by Na-K ATPase.

Tonicity describes the effect of solutions on cell volume. If cells are placed in a solution with low osmolarity, water enters the cells and they swell. Such solutions are called *hypotonic*. In solutions of high osmolarity, water leaves the cells and they shrink. These solutions are *hypertonic*. Body fluids are normally isotonic, but fluids may temporarily shift between ECF and ICF if there is an osmotic imbalance.

Fluid Shifts and Blood Volume

Explain how body water compartments shift with changes in volume and osmolarity

Blood pressure depends on cardiac output, peripheral resistance, and blood volume. Increased *blood volume* increases blood pressure. Several conditions can alter ECF volume, resulting in either increases or decreases in blood pressure. All of these are the result of altered *sodium balance*. If Na intake is greater than Na excretion by the kidneys, extra Na is retained in the body and the ECF volume increases, causing an increase in blood pressure and possibly accumulation of fluids in the tissues (*edema*). When Na excretion is greater than Na intake, the ECF decreases (*volume contraction*) and blood pressure drops.

> Interstitial fluids and plasma are in free exchange (one compartment)
> Isosmotic volume loss decreases ECF volume with no effect on ICF
> Loss of dilute salt and salt ingestion are hyperosmotic: ICF shifts to ECF

Consider three examples. 1) *Diarrhea* causes the loss of isosmotic fluids. The osmolarity of the lost fluid is the same as that of the ECF. Because the osmolarity has not changed, there is no fluid shift between ECF and ICF. The loss of fluids simply decreases blood volume (*isosmotic volume contraction*) and blood pressure. 2) *Sweating* causes the loss of a dilute solution of water and NaCl. More water than salt is lost, so sweating decreases the volume and increases the osmolarity of the ECF (making it hyperosmotic). Fluids shift from ICF to ECF, depleting cells of water and increasing their osmolarity, but this shift is partial and ECF volume is lower than normal (*hyperosmotic volume contraction*) leading to a drop in blood pressure. 3) Dry salt ingestion, such as by eating a bag of potato chips, increases ECF osmolarity as a result of the increased salt concentration. Water shifts from ICF to ECF to balance the increased ECF osmolarity. ECF volume increases (*hyperosmotic volume expansion*), as does blood pressure.

Water Balance

Describe how body water balance is maintained

Water input equals water output to remain in balance. *Water balance* is required to maintain stable blood volume and ionic composition. Increased perspiration and breathing during exercise and in dry environments can cause significant water loss. Up to 2 L per hr can be lost in sweat. If this is not made up, venous return is reduced which reduces stroke volume and may compromise tissue perfusion due to decreased cardiac output. Increased heart rate helps maintain cardiac output and a reflex reduction in skin blood flow limits heat loss, which increases heat load. In addition, sweating causes loss of electrolytes which must be replaced (beer was prescribed for miners in the 1920s to maintain water and electrolyte balance).

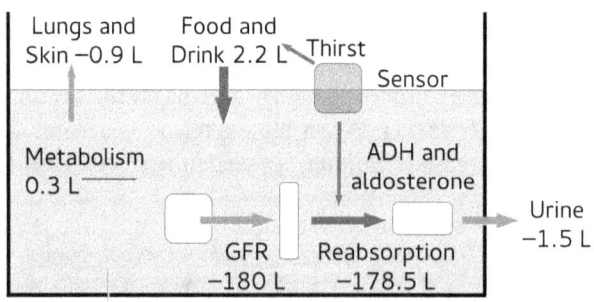

Under ordinary conditions, most water loss is due to production of urine. This body water loss (*dehydration*) can only be made up by water from outside the body. Reabsorption of water in renal distal tubules makes urine more concentrated and conserves body water, but cannot replace water loss. Dehydration sends signals to higher brain centers that cause conscious craving of water (*thirst*). *Osmotic thirst* is due to dehydration (loss of water) which increases plasma osmolarity. Hypothalamic osmoreceptors activate water seeking and thirst. Salt ingestion increases plasma osmolarity, conserving water and increasing thirst, thus increasing blood volume and blood pressure.

> ADH release by posterior pituitary gland is stimulated by
> increased osmolarity of ECF sensed by hypothalamic osmoreceptors
> ADH increases water reabsorption and hypothalamus stimulates thirst

Volumetric thirst results from decreased blood volume, which is often due to blood loss, vomiting, or diarrhea, and is sensed by baroreceptors which activate the renin-angiotensin system to produce angiotensin II that acts on the hypothalamus to induce thirst, causes vasoconstriction to restore blood pressure, and decreases water loss in the urine. Rehydration is limited by the gastric emptying rate to about 1 L per hr with ideal fluids.

Antidiuretic Hormone

Describe the mechanism by which ADH release is regulated

Water is conserved by increasing water reabsorption in kidneys. *Antidiuretic hormone* (ADH, vasopressin) inserts aquaporin water channels in distal tubules and collecting ducts to allow osmosis. In the collecting ducts osmosis is driven by the medullary osmotic gradient established by the loop of Henle. In the distal convoluted tubules, *aldosterone* is required to increase salt reabsorption that drives water reabsorption. In both cases, without ADH, the distal tubules are impermeable to water. So, ADH is required to allow osmosis which results in a more concentrated urine. With extreme fluid loss and drop in blood pressure, glomerular filtration rate (GFR) can slow to conserve water, but within the ordinary range of blood pressures, GFR remains constant.

> ADH is produced in neurosecretory cells of hypothalamus which project
> into posterior pituitary gland where they are released into blood

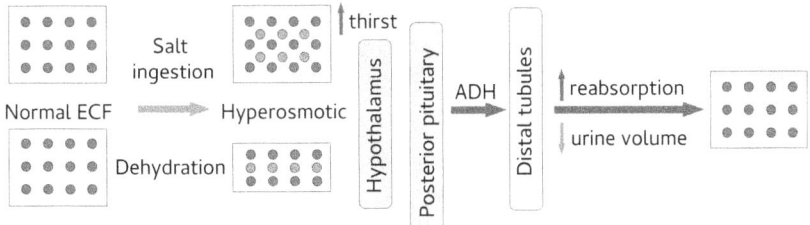

ADH is primarily released in response to increased *plasma osmolarity* (dehydration), which is sensed by osmoreceptors in the hypothalamus.

Neurosecretory cells that arise in the hypothalamus project into the posterior pituitary gland, where the hormone is released to enter the bloodstream. The name *vasopressin* comes from its vasopressor activity at high concentrations, that contributes to restoring blood pressure in hypovolemic shock. The signal for release of high levels of vasopressin is mediated by stretch receptors (cardiopulmonary baroreceptors) within the atria and large veins. ADH is also released in a sympathetic response to decreased blood pressure and angiotensin II production by the renin-angiotensin system.

Renin-Angiotensin-Aldosterone System

Explain the role of the RAAS in maintaining blood pressure

Decreased blood pressure (or volume) stimulates *juxtaglomerular cells* between macula densa and glomerulus to secrete renin. *Renin* activates a cascade of reactions that leads to production of the hormones angiotensin II and aldosterone (*renin-angiotensin-aldosterone system,* or RAAS). Renin is an enzyme that converts inactive hepatic *angiotensinogen* to angiotensin I. ACE (*angiotensin-converting enzyme*) in lungs and other capillaries activates angiotensin I to *angiotensin II.*

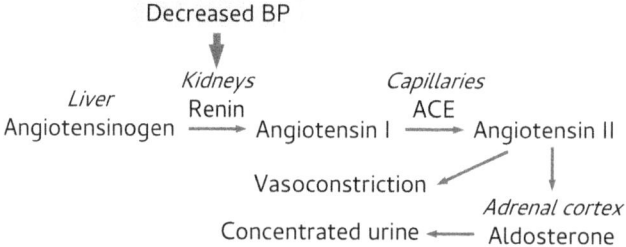

Angiotensin II causes vasoconstriction and secretion of aldosterone from the adrenal cortex. *Vasoconstriction* minimizes blood loss in bleeding and increases blood pressure. *Aldosterone* inserts Na pumps and transporters in distal tubules and collecting ducts. Thus, Na is reabsorbed at the expense of K secretion. Water reabsorption follows Na by osmosis (when ADH is present), producing a concentrated urine and preserving body water.

Maintenance of body water ensures that blood volume and blood pressure do not decrease, but does not actively increase blood volume. Replacement of body water with ingested liquids leads to a higher blood volume than in the absence of aldosterone, causing increased blood pressure. *ACE inhibitors* (*lisinopril*) and *angiotensin II receptor blockers* (ARBs, such as *losartan*) are commonly used to treat hypertension.

Atrial Natriuretic Peptide

Describe the source and actions of atrial natriuretic peptide

ADH and aldosterone are antidiuretic hormones. In response to increased blood osmolarity or decreased blood pressure (or both), they increase the volume of water reabsorbed in the kidneys, producing a concentrated urine. *Diuresis* is increased urine volume, which results in production of a diluted urine. *Diuretics* reduce blood volume and pressure by decreasing the amount of water reabsorbed by the kidneys which produce a copious urine. Diuretic medications block Na reuptake pumps in the renal tubules or block the production of aldosterone.

Atrial natriuretic peptide (ANP) is a natural diuretic produced by the atria when they are stretched due to increased blood volume and venous return pressure. They have effects opposite to those of aldosterone. ANP decreases Na reabsorption in the distal tubules and collecting ducts. This reduces water reabsorption, which decreases blood volume and pressure.

Stretch of atria in hypertension produces ANP that increases GFR
and reduces Na reabsorption in distal tubules and collecting ducts
Diuresis produces dilute urine and reduces blood volume and pressure

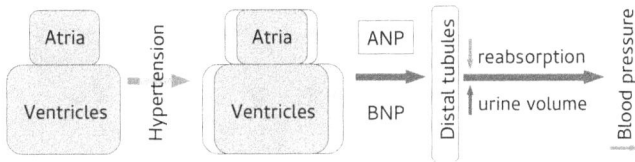

The effects of ANP supplement the decreased aldosterone production in response to lowered blood pressure through the RAAS and directly inhibit renin secretion. In addition, ANP dilates afferent arterioles and constricts efferent arterioles to maintain GFR in the face of reduced systemic blood pressure, complementing the effects of vasopressors released from the macula densa. Brain natriuretic peptide (BNP) is secreted by ventricular myocytes in response to stretch, but has a tenfold lower activity than ANP and is not the primary agent responsible for maintaining blood volume in hypotension. However, its long half-life makes it better suited as a vascular marker of heart failure.

Potassium Balance

Explain how K ion balance is maintained and why it is important

In the proximal convoluted tubules, K ions are reabsorbed along with Na ions and water. In the ascending limb of the loop of Henle, K ions enter the cell in exchange for Na and Cl ions by the NKCC2 pump. K ions are then free to diffuse down their electrochemical gradient either into the lumen or back into the interstitial fluids. Both of these processes are unregulated.

K ion levels in the ECF are closely regulated by balancing ingestion and excretion in the kidneys. The distribution of K ions across cell membranes is a primary determinant of membrane potential. Thus, K imbalances alter neuronal excitability and muscle contraction. *Hypokalemia* (low K ion levels in the blood) causes muscle weakness that may result in respiratory or cardiac failure. *Hyperkalemia* causes life-threatening cardiac arrhythmias.

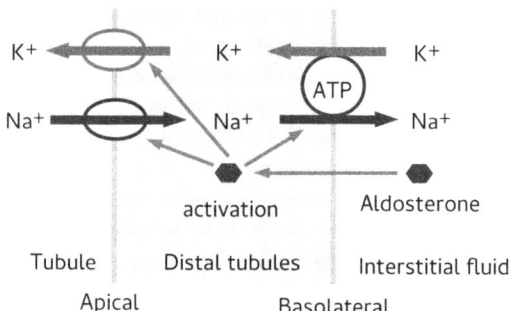

Aldosterone activates Na and K channels and Na pump in principal cells of distal tubules which increases flux of both ions (Na reabsorption and K secretion)

With a high potassium diet, intracellular K ion concentrations in the principal cells of the distal tubules increases, resulting in enhanced secretion of K ions to maintain ionic balance. With a low potassium diet, intracellular K decreases and more is reabsorbed in the distal tubules. Aldosterone is secreted from the adrenal cortex directly in response to elevated ECF K ion levels. Aldosterone increases secretion of K ions in the principal cells of distal tubules by increasing Na reabsorption. Activity of Na-K ATPase and the number of Na channels in the luminal membrane are both increased. The result is an increase in intracellular K ions. At the same time, aldosterone increases K ion channels in the luminal membrane to ensure that the increased intracellular K ions are secreted into the lumen.

K ion levels in principal cells are also influenced by pH. A basolateral H-K ATPase responds to acidosis by pumping H ions into the cell in exchange for K ions. Thus, acidosis increases K ion reabsorption, leading to hyperkalemia. In alkalosis, H ions leave the cells, while K ions enter the luminal cells and is secreted. This may cause hypokalemia.

Calcium Balance

Explain the role of the kidneys in maintaining Ca balance

ECF total calcium is regulated at 2.5 mM; cellular calcium levels are very low (submicromolar). *Parathyroid hormone* (PTH) is secreted by the parathyroid glands in response to decreased blood calcium. PTH mobilizes calcium from bones, increases renal calcium reabsorption, and stimulates dietary Ca absorption. Renal *calcitriol* increases the uptake of Ca from the diet. Dietary vitamin D is converted by the liver and then in the kidney to make calcitriol. Kidney calcitriol production is increased by prolactin in lactating mothers.

In proximal tubules and ascending limb, paracellular Ca reabsorption is driven by Na gradient established by Na pump and luminal Na channels
In ascending limb Ca reabsorption is aided by paracellular claudin

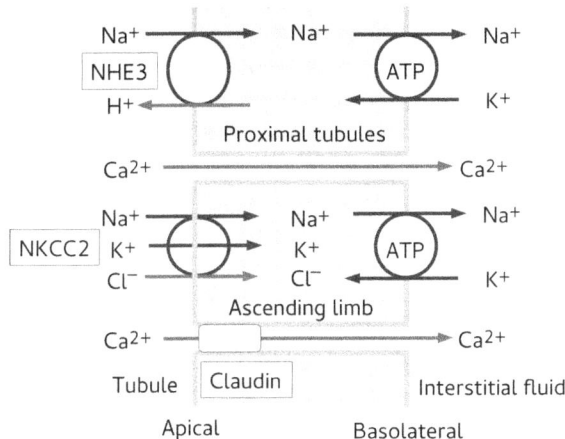

Most Ca is reabsorbed in *proximal convoluted tubules*. A significant amount is reabsorbed in the ascending limb of the loop of Henle, through a paracellular route driven by Na reabsorption. Thus, loop diuretics not only reduce Na reabsorption, they reduce Ca reabsorption, as well. For this reason, they can be used to treat hypercalcemia.

PTH also inhibits the luminal Na-phosphate cotransporter in the PCT, which decreases phosphate reabsorption. Thus, PTH increases plasma

Ca concentrations and decreases phosphate concentrations to prevent formation of Ca-phosphate salts.

A significant amount of calcium in blood is bound to proteins; only about 60% is available to be filtered in the renal corpuscle. Calcium is reabsorbed with Na ions in the PCT and thick ascending limb of the loop of Henle. In both regions, Ca transport is paracellular. Thus, loop diuretics inhibit Ca reabsorption along with Na reabsorption.

In *distal tubules*, Ca reabsorption is hormone-dependent. It is not coupled to Na reabsorption. In the distal tubules, parathyroid hormone (PTH) increases activity of an apical Ca channel (TRPV5). Ca is extruded from the tubular epithelial cells by a Ca-ATPase and sodium-calcium exchanger (NCX1). Thiazide diuretics spare Ca by an unknown mechanism.

> Ca enters distal tubule epithelial cells through apical TRPV5 channel
> PTH stimulates production and activates TRPV5
> Ca is pumped out by Ca-ATPase and leaves through Na-Ca exchanger

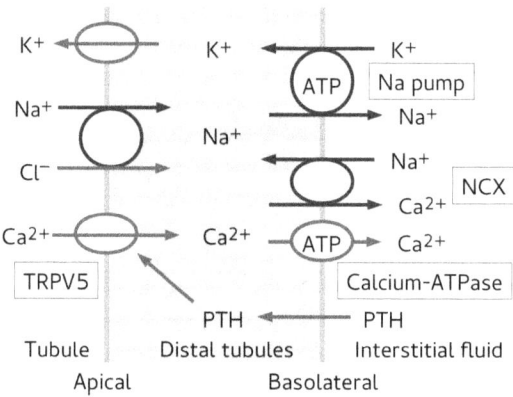

Calcium is important for exocytosis, muscle contraction, and neural function. Hypocalcemia is low blood calcium. It makes the nervous system hyperexcitable. In extreme cases, tetany of respiratory muscles can cause asphyxiation. Hypercalcemia is elevated blood calcium. This depresses neuromuscular activity.

Renal Regulation of pH

Explain how the kidneys regulate bicarbonate concentrations and help maintain constant pH

Only a narrow range of pH is compatible with life (7.35-7.45). *Acids* produce hydrogen ions and have pH < 7. *Bases* take up hydrogen ions and have pH > 7. *Buffers* keep the pH from changing when acid or

base is added to a solution. In the ECF, bicarbonate is the primary buffer system. Plasma bicarbonate is the primary transport form of carbon
dioxide in blood. Thus, changes in rate and depth of breathing produce
rapid adjustments to blood pH by changing bicarbonate concentrations.
Increased circulating carbon dioxide (decreased ventilation) decreases
pH. Reduced carbon dioxide (increased ventilation) increases pH. In
addition, kidneys directly regulate both pH and bicarbonate levels by
controlling their secretion and reabsorption.

Filtered bicarbonate converted to carbon dioxide by carbonic anhydrase
enters proximal tubule cells to be reabsorbed by bicarbonate transporters
Acids (H+) are secreted by luminal NHE exchanger

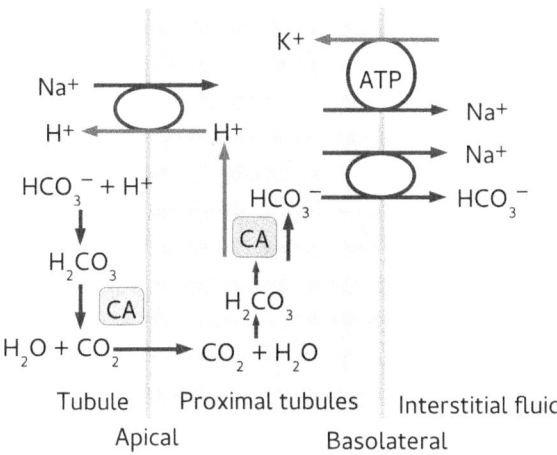

The *proximal convoluted tubules* secrete acids and reabsorb bicarbonate to rid the body of acids produced by metabolism. Filtered
bicarbonate is converted to carbon dioxide by luminal carbonic anhydrase (CA). Carbon dioxide diffuses into cells where it is reconverted
to bicarbonate by the intracellular form of the same enzyme. This bicarbonate is reabsorbed through bicarbonate transporters on the
basolateral membrane. Acids (H+) produced by carbonic anhydrase are
secreted back into the tubular lumen by a Na-H exchanger (NHE). This
on-going process rids the body of metabolic acids while maintaining a
bicarbonate reserve in plasma that buffers acids produced by the body.
To maintain normal pH. this bicarbonate reserve must be twenty-fold
higher than the carbonic acid concentration in plasma (20:1).

Plasma carbon dioxide enters type A intercalated cells of distal tubules
Bicarbonate produced by carbonic anhydrase is reabsorbed
Acids (H+) are secreted by luminal H-K ATPase (proton pump)

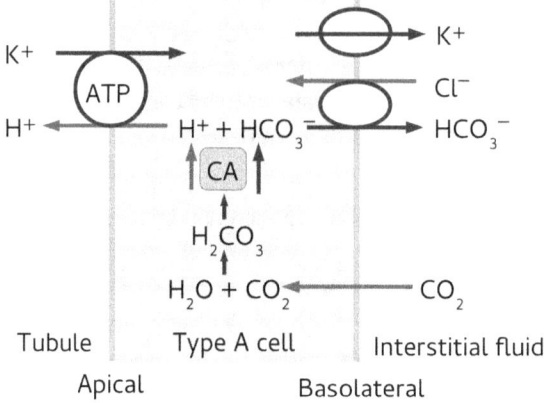

Distal tubules can either reabsorb or secrete bicarbonate to maintain blood pH. The mechanism is similar to that in the proximal tubules and depends on a high activity of carbonic anhydrase in the distal tubule cells. Carbon dioxide enters the cell where it is converted to H+ and bicarbonate by carbonic anhydrase. Protons are pumped out of the cell by a H-K ATPase (proton pump) and bicarbonate leaves through a chloride ion antiporter.

Two types of cells in the distal tubules carry out this function. *Type A intercalated cells* respond to acidosis by secreting H^+ and reabsorbing bicarbonate. *Type B intercalated cells* respond to alkalosis by secreting bicarbonate and reabsorbing acids. The transporters are the same, but they are switched so that luminal transporters of type A cells are on the basolateral surface of type B cells and vice versa. Acidosis is associated with hyperkalemia because the increased pumping of H ions is associated with increased uptake of K ions through the luminal proton pump.

Acidosis and Alkalosis

Define acidosis and alkalosis and mechanisms of compensation

Acidosis is blood pH < 7.35. *Alkalosis* is defined as a blood pH > 7.45. With a bicarbonate buffer system, the ratio of bicarbonate to carbonic acid must be 20:1 to maintain a pH of 7.4. An acid-base imbalance can be the result of changes in either component of the buffer system. If it is due to bicarbonate, the imbalance is called *metabolic*. If it is due to elevated or depressed carbonic acid levels, it is *respiratory*, since carbonic acid is produced from carbon dioxide by combining with water.

Bicarbonate and carbonic acid levels are reported in measurements of *arterial blood gases* (ABG). Bicarbonate is measured in mM and carbonic acid is reported as $PaCO_2$. In analyzing an acid-base balance, begin with pH. If pH < 7.35, it is acidosis; if pH > 7.45, it is alkalosis. Then, look at the imbalance in bicarbonate and $PaCO_2$ to distinguish metabolic from respiratory dysfunction. The normal range for bicarbonate is 22 to 26 mEq/L and that for $PaCO_2$ is 35 to 45 mm Hg. If the level of bicarbonate is outside the normal range and in the direction expected for the acid-base imbalance, it is a metabolic. If the $PaCO_2$ is outside the normal range and in the direction expected, it is respiratory.

1) Low bicarbonate reduces pH. Thus, pH < 7.35 with bicarbonate < 22 mEq/L is *metabolic acidosis*.

2) High bicarbonate increases pH. Thus, pH > 7.45 with bicarbonate > 26 mEq/L is *metabolic alkalosis*.

3) High CO2 reduces pH. Thus, pH < 7.35 with $PaCO_2$ > 45 mm Hg is *respiratory acidosis*.

4) Low CO2 increases pH. Thus, pH > 7.45 with $PaCO_2$ < 35 mm Hg is *respiratory alkalosis*.

The most common acid/base imbalance is acidosis. In acidosis, the kidneys excrete H ions to rid the body of acids and reabsorb bicarbonate to add to the buffer pool. Both serve to increase pH and reverse the acidosis. In lactic acidosis, poor oxygen supply cannot keep up with anaerobic lactate production. In ketoacidosis, often associated with uncontrolled type 1 diabetes and low carbohydrate diets, ketone bodies are produced from amino acids. Bicarbonate may be lost by diarrhea, which also causes metabolic acidosis.

Drug-induced hypoventilation, increased airway resistance in asthma, and impaired gas exchange in COPD are common causes of respiratory acidosis. Metabolic alkalosis may result from excessive vomiting or overuse of bicarbonate-containing antacids. Anxiety may cause hyperventilation that leads to a temporary respiratory alkalosis.

Compensation

Explain the mechanisms of respiratory and renal compensation for acid-base imbalances

The body compensates for acid-base disturbances by activating the opposite system. In metabolic disturbances, compensation is respiratory; in respiratory disturbances, compensation is renal. Metabolic disturbances are always compensated. Respiratory disturbances may be uncompensated (acute) or compensated (chronic). *Compensation* may partly restore pH levels to near normal but has minimal or no effect on the primary disturbance.

In metabolic acidosis, *hyperventilation* reduces $PaCO_2$ levels, decreasing carbonic acid, and partly restoring pH to normal values. Some long-term renal compensation may also occur, but increased respiration is a common sign of metabolic acidosis. The compensatory response to metabolic alkalosis is hypoventilation and increased $PaCO_2$.

In respiratory acidosis, the kidneys secrete H ions and reabsorb bicarbonate. Thus plasma bicarbonate levels are increased when $PaCO_2$ increases in chronic (compensated) respiratory acidosis. The opposite effect is observed in respiratory alkalosis. The kidneys secrete bicarbonate and reabsorb H ions. Plasma bicarbonate decreases along with the decrease in $PaCO_2$ in chronic respiratory alkalosis.

These six possibilities can be summarized in a table. When compensated, bicarbonate and $PaCO_2$ levels go in the same direction: both either increase or decrease. If both decrease and pH is slightly acidic it is metabolic acidosis; if pH is slightly alkaline it is compensated respiratory alkalosis. Conversely, if both increase and pH is slightly alkaline it is metabolic alkalosis; if slightly acidic, it is compensated respiratory acidosis.

Condition	pH	Disturbance	Compensation
Metabolic acidosis	< 7.35	Bicarb < 22 mEq/L	↓ $PaCO_2$
Metabolic alkalosis	> 7.45	Bicarb > 26 mEq/L	↑ $PaCO_2$
Acute Respiratory Acidosis	< 7.35	$PaCO_2$ > 45 mm Hg	None
Chronic Respiratory Acidosis	< 7.35	$PaCO_2$ > 45 mm Hg	↑ Bicarbonate
Acute Respiratory Alkalosis	> 7.45	$PaCO_2$ < 35 mm Hg	None
Chronic Respiratory Alkalosis	> 7.45	$PaCO_2$ < 35 mm Hg	↓ Bicarbonate

Review Questions

1. What are the three major body fluid compartments? How are they separated from each other? What are their volumes?
2. How does osmolarity of ECF determine whether intracellular fluids will shift out of or into cells?
3. What are the inputs and outputs for body water? Which of these are regulated?
4. How does antidiuretic hormone (ADH) increase water reabsorption in the kidney?

5. How does activation of the renin-angiotensin system lead to aldosterone secretion?
6. What are the effects of angiotensin II?
7. What is the primary source of ANP and what are its effects?
8. What is the effect of aldosterone on K secretion in the distal tubules?
9. How is Ca reabsorbed in proximal tubules and ascending limb of the loop of Henle?
10. How is bicarbonate reabsorbed in the proximal tubules?
11. How is bicarbonate reabsorbed in distal tubules?
12. How do type A and type B intercalated cells differ?
13. How are acidosis and alkalosis defined?
14. How can metabolic imbalances be distinguished from respiratory imbalances?
13. How does compensation help restore blood pH?

Digestion and Metabolism

My opinions may be doubted, denied, or approved, according as they conflict or agree with the opinions of each individual who may read them; but their worth will be best determined by the foundation on which they rest—the incontrovertible facts.

– William Beaumont

Digestive Tract

Kidney nephrons illustrate many of the most fundamental principles of physiology we have learned so far. It maintains *homeostasis* of plasma osmolarity by removing wastes, adjusting Na ion concentrations, and preserving long-term blood pH nearly constant. Each region of the nephron has distinct functions, showing the importance of compartmentation of functions. These functions are carried out by distinct transporters and channels in the membranes of the cells surrounding the tubules. *Compartmentation* of functions is due to compartmentation of the *proteins* that carry out those functions. Finally, the nephron demonstrates the close relationship between *blood* and systemic functions. The heart pumps blood, the lungs exchange waste carbon dioxide for fresh oxygen in the blood, and the kidneys remove wastes from the blood.

Now, we turn to the digestive system, which provides *nutrients* to cells by digesting foods to molecules that can be absorbed and carried in the blood. The digestive system extracts as much nutrition as possible from any meal. Homeostasis of nutrient intake is determined by *appetite control*. What small amount of food taken in that cannot be digested is excreted from the body as *feces*.

Like the nephron tubule, the *digestive tract* is a tube with distinct functions that vary along its length. Digestion is carried out by *enzymes* that hydrolyze (add water to chemical bonds) of macronutrients. *Carbohydrate digestion* begins in the mouth with salivary amylase that begins breakdown of starches. *Protein digestion* begins in the mouth where gastric acid unfolds proteins that then are hydrolyzed by pepsin. Fats are not digested to any significant extent until they reach the duodenum where *bile salts* from the liver emulsify fat droplets into tiny micelles that can be acted on by *pancreatic lipase*.

Digestive System

Describe the overall organization of the digestive system

In the *digestive system*, organs break down foods, absorb nutrients, and excrete solid wastes. The *digestive tract* forms a single tube, divided into structural and functional compartments. It includes the mouth, pharynx, esophagus, stomach, small and large intestines, and rectum.

Accessory organs produce materials that enter the digestive tract to aid in digestion. The accessory organs are the salivary glands, liver, gall-bladder, and pancreas.

> Digestive system breaks down foods, absorbs nutrients, secretes wastes
> Digestive tract: single tube from mouth to rectum
> Accessory organs secrete materials into digestive tract to aid digestion

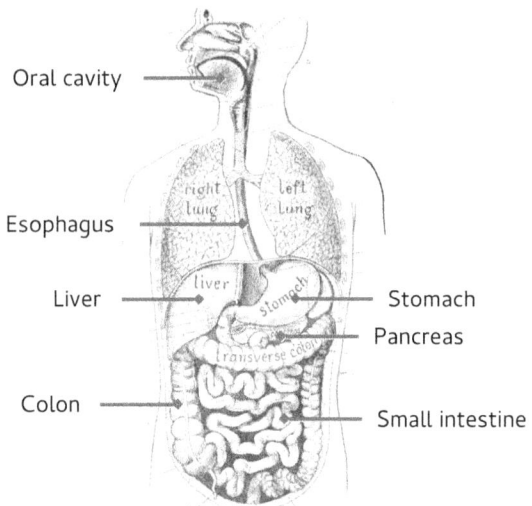

Digestion is the breakdown of *macronutrients* (carbohydrates, proteins, and fats) to simple *nutrients* that can be absorbed. It is divided into two phases that go on together in each compartment of the digestive tract. *Mechanical digestion* is the physical breakdown of foods. This includes chewing (mastication) and mixing in the stomach and small intestine. *Chemical digestion* is the breakdown of macronutrients by enzymes. Most of this breakdown occurs by enzymatic addition of water across chemical bonds (*hydrolysis*).

Once fully digested, nutrients can be absorbed. *Absorption* is the uptake of simple nutrients into the capillaries and lymphatics (lacteals) of intestinal villi. Those food materials that are ingested but not digested and absorbed are defecated through the anus (*defecation*).

Mastication

What is mastication and what are the main components of saliva?

In the oral cavity, *mastication* (chewing) breaks up food particles. The teeth and tongue play a major role in this process. The *salivary glands* secrete a watery fluid that mixes with foods to make them easier to swallow. Taste receptors on the tongue require tastants to be dissolved in order to be detected. *Saliva* ensures that the oral cavity does not dry out. Salivation is mostly under parasympathetic control by cranial nerves VII and IX.

> Mastication (chewing) breaks up food particles using teeth and tongue
> Salivary glands secrete watery fluid that moistens food to form a bolus
> Saliva contains amylase that begins digestion of starches

Salivary amylase begins digestion of complex carbohydrates (mostly starches) to disaccharides in the oral cavity. This digestion is stopped by acids in the stomach and continues in the small intestine where pancreatic amylase is secreted. Once the food has been thoroughly chewed, mixed with saliva, and partially digested, it forms a *bolus*, which is soft and can readily be swallowed.

GI Tract Layers

Describe the four layers of the gastrointestinal wall

The wall of the digestive tract consists of four tissue layers surrounding a hollow lumen. The innermost layer of the digestive tract is the *mucosa*. In the oral cavity and esophagus, the mucosa is abrasion-resistant stratified squamous epithelium. From the stomach to the anus, the mucosa is simple columnar epithelium.

> Mucosa: innermost epithelial layer of digestive tract with ducts and glands
> Muscularis has two or three muscle layers innervated by myenteric plexus
> Serosa (serosal membrane) or adventitia is outermost layer

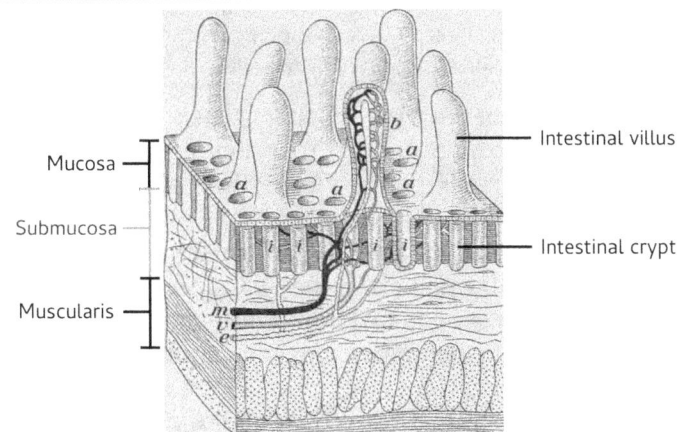

The *submucosa* is connective tissue that lies deep to the mucosa. The *muscularis* (muscularis externa) has inner circular and outer longitudinal layers of muscle throughout the digestive tract. The stomach has a third, oblique layer of muscle in its muscularis that aids in mixing. The outermost layer is an *adventitia* that attaches to surrounding tissues or a well-defined epithelial *serosa*.

The *myenteric plexus*, which is part of the enteric nervous system, innervates muscle layers in the stomach and throughout the gastrointestinal (GI) tract. These nerves arise in the medulla oblongata and pass to the stomach through the vagus nerve (CN X).

ANS in Gastrointestinal Tract

Describe the innervation of the GI tract by the autonomic nervous system

Two networks of neurons make up the autonomic supply of the digestive system. The *myenteric (Auerbach's) plexus* lies between the circular and longitudinal muscle layers of the the muscularis externa, the outer muscular layer of the digestive tract (from upper esophagus to internal anal sphincter). This layer is responsible for propulsion (peristalsis) and mixing (segmentation) of the contents of the GI tract.

The *submucosal (Meissner's) plexus* is in the connective tissue of the submucosal layer of the intestine (not esophagus or stomach). This layer lies deep to the mucosal inner lining of the digestive lumen. Here, parasympathetic preganglionic neurons synapse with ganglionic fibers that supply the *muscularis mucosa*, a thin layer of mucosal smooth muscle cells adjacent to the submucosa. They stimulate secretion from glands of the submucosal layer into the digestive tract.

> Autonomic nervous system innervates gastrointestinal (GI) tract
> Myenteric plexus within muscularis externa controls GI wall contraction
> Submucosal plexus controls GI tract secretions from its glands

Sensory neurons in the GI tract monitor stretch (mechanoreceptors) and the contents of the digestive tube (chemoreceptors and osmoreceptors). For example, *stretch receptors* in the wall of the upper stomach signal the presence of food. These signals travel to the CNS via the *vagus nerve* where a parasympathetic signal is returned to the myenteric plexus of the stomach, again along the vagus nerve. Since the vagus nerve carries both sensory and motor information for the response, it is a *vagovagal reflex*. The neurotransmitter is the peptide, VIP (vasoactive intestinal peptide) which causes relaxation of smooth muscle cells in the walls of the stomach. This receptive relaxation allows the upper stomach to stretch as it receives food.

Parasympathetic neurons of the GI tract release either acetylcholine (cholinergic neurons) or peptides, such as VIP or substance P (peptidergic neurons). Their ganglia are within the myenteric and submucosal plexuses. Sympathetic nerves synapse in collateral ganglia close to, but outside, the GI tract. All sympathetic neurons innervating the GI tract secrete norepinephrine. In the lower stomach, parasympathetic innervation increases gastric contractions while sympathetic stimulation decreases the force of contractions.

Enteric Nervous System

Describe the organization and functions of the enteric nervous system

The *enteric nervous system* (ENS) is a set of neurons and ganglia in the gastrointestinal (GI) tract that directly control motility and secretion. The neurons and autonomic ganglia of the enteric nervous system lie wholly within the myenteric and submucosal plexuses. GI tract ganglia also communicate with each other through interneurons and the collateral ganglia of the sympathetic division. The CNS dominates control of contractile activity in the esophagus and stomach, acid secretion, and inhibition of defecation. The small intestine and colon are largely controlled by the ENS. Although it receives input from sympathetic and parasympathetic neurons, the enteric nervous system can operate nearly independently of the brain and spinal cord.

> Enteric nervous system: group of ganglia and neurons within GI plexuses
> ENS communicates with sympathetic collateral ganglia and CNS
> Sympathetic and parasympathetic innervation modulates ENS activity

Neurotransmitters of the enteric nervous system are as various as those within the brain, including some two dozen neuropeptides. Serotonin and dopamine are more abundant in the gut than in the brain. ENS ganglia mediate local (*short*) reflexes. These are distinct from the *long reflexes* that are integrated in the CNS. Short reflexes maintain motility in the small and large intestines by controlling activity of excitatory and inhibitory neurons that supply the smooth muscle cells of the intestinal wall. In addition, enteric reflexes regulate fluid and electrolyte movement across the intestinal wall and mucosal (local) blood flow.

> Long reflexes are integrated in CNS (brain and spinal cord)
> Short reflexes bypass CNS and are integrated in local autonomic ganglia

Swallowing

Describe the process of swallowing

Swallowing (*deglutition*) is initiated voluntarily. It is divided into three phases: buccal, pharyngeal, and esophageal. In the *buccal phase*, the tongue pushes the bolus against the soft palate and back of the mouth.

The *soft palate* (the posterior part of the upper mouth) elevates to seal off the *nasopharynx* that connects the oral cavity to the nasal cavity and the swallowing reflex is triggered in the medulla oblongata. In the *pharyngeal phase*, breathing is inhibited in the brainstem respiratory center and the *epiglottis* closes to block the trachea. In the esophageal phase, food is propelled down the esophagus by peristalsis.

> Swallowing (deglutition) begins when bolus is pushed to back of mouth
> Soft palate elevates, swallowing reflex is triggered, respiration is inhibited
> Epiglottis closes as bolus passes down tube by peristalsis

In *peristalsis*, circular muscle in the esophageal wall contracts behind the bolus while the longitudinal muscle relaxes, forming a constriction behind the bolus. In front of the bolus, circular muscle relaxes to receive the bolus. The *lower esophageal sphincter* at the entrance to the stomach opens to allow the bolus to pass. It closes once the bolus enters the stomach.

> Peristalsis pushes materials along digestive tube
> Circular muscle contracts behind bolus to push it into
> distal region where circular muscle is relaxed

Stomach Anatomy

Describe the structure of the stomach muscle layers and process of gastric mixing

Foods in the mouth are formed into a compact bolus by the action of the tongue and mastication. The bolus is swallowed and passes down the esophagus to enter the stomach at its cardia. The stomach has folds (*rugae*) that allow it to expand when filled with food.

> Food passes into cardia of stomach through lower esophageal sphincter
> Longitudinal, circular, and oblique muscle layers churn and mix foods
> with gastric gland secretions prior to emptying through pyloric sphincter

The stomach has three muscle layers: longitudinal, circular, and oblique. The oblique muscle layer, which is only found in the stomach, contributes to churning contractions. Peristalsis begins in the fundus (the upper part of the stomach) and passes through the body to exert pressure on the *pyloric sphincter* at the entrance to the duodenum. While the sphincter is closed, stomach contents reflux back to mix and grind the food. This sphincter opens slightly to allow the liquid chyme formed in the stomach to pass in a controlled (slow) fashion into the first part of the small intestine (the duodenum).

> Stomach contents are mixed by peristaltic wave originating in fundus
> Contents reflux when pyloric sphincter is closed to make liquid chyme
> that is released slowly into duodenum to limit amount to be digested

Gastric Glands

Describe the cell types of gastric glands and their secretions

In the stomach wall, *gastric glands* contain secretory cells. The open-ing to the glands is the *gastric pit*. One set of cells (parietal cells, mucous neck cells, and chief cells) secrete chemicals that continue the process of digestion and protect the stomach lining. *Parietal cells* se-crete HCl (*gastric acid*) and *intrinsic factor*, which complexes vitamin B12. Gastric acid unfolds (denatures) proteins, exposing their peptide bonds to proteases that cleave the peptide bonds that hold proteins to-gether. *Mucous neck cells* secrete bicarbonate and mucus, forming a layer that protects the stomach lining. *Chief cells* secrete pepsinogen, which is converted to active pepsin by acid hydrolysis. Pepsin begins hydrolysis of denatured proteins.

> Gastric glands: invaginations of stomach wall containing secretory cells
> Gastric pit is entrance to gastric glands

Another set of cells (G cells and ECL cells) secrete hormones that reg-ulate stomach activity. *G cells*, which are concentrated in the antrum and pylorus at the exit of the stomach, produce gastrin. *ECL cells* (en-terochromaffin-like) secrete histamine that acts as a paracrine to locally stimulate HCl secretion.

> Parietal cells secrete gastric acid; chief cells secrete pepsinogen
> Mucous neck cells secrete mucus and bicarbonate

Control of Gastric Activity

Describe the three phases of gastric activity

Gastric activity is controlled in three phases: cephalic, gastric, and in-testinal. In the *cephalic phase*, the sight, smell, and taste of food increases gastric secretions to prepare the stomach for incoming food. *Gastrin* is released by the stomach due to distension and the presence of foods. *Histamine* is released locally by ECL cells. Both increase gastric acid secretions.

> Smell, sight, taste, thoughts of food initiate gastric activity (cephalic phase)
> Food in stomach (gastric phase) acts on stretch and chemoreceptors
> Reflex stimulates secretion of gastric acid, mucous, pepsinogen, and gastrin

In the *gastric phase*, food in the stomach triggers reflexes to increase its secretions and motility. Gastrin is released into the bloodstream due to distension of the stomach (sensed by stretch receptors) and the presence of foods (sensed by chemoreceptors). Gastrin stimulates cells of the stomach to secrete gastric acid. Activation of the parasympathetic vagus nerve further stimulates gastric motility and secretions.

Somatostatin, released in the pyloric antrum (the entrance to the pyloric sphincter), decreases gastric acid secretions as chyme enters the duodenum. In the *intestinal phase* of the control of gastric activity, food in the duodenum shuts down stomach secretions and motility. The presence of lipids and carbohydrates stimulate secretion of CCK (*cholecystokinin*). Decreased pH in the duodenum stimulates *secretin* release. CCK and secretin inhibit the secretion of gastric acid.

> Presence of foods in duodenum produces hormones that
> decrease stomach secretions and motility (intestinal phase)

Liver and Gallbladder

Describe the structure of the liver, gallbladder, and bile duct system

The *liver* is inferior to the diaphragm. It is largest on the right side of the body. The right and left lobes are separated by the falciform ligament. The falciform ligament stabilizes the position of the liver relative to the diaphragm and abdominal wall. The caudate (superior) and quadrate (inferior) lobes are in the posterior midline. Blood vessels and ducts are in-between these two small, posterior lobes.

> Liver is inferior to diaphragm and superior to all other digestive organs
> Gallbladder lies in pocket on posterior surface
> Pancreas is dorsal (lies behind) stomach

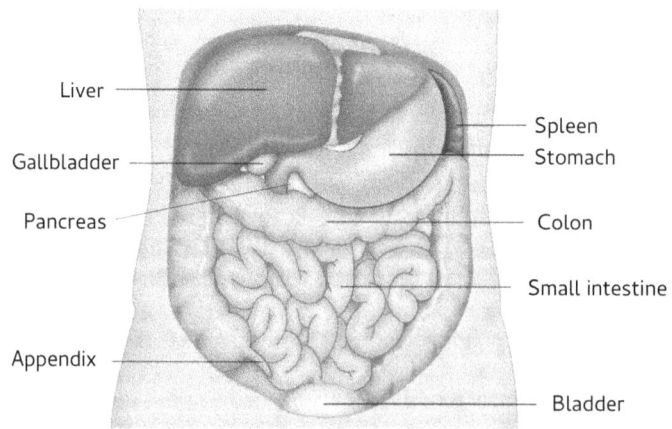

Bile secreted by the liver empties into bile ductules that join to form the bile ducts of the hepatic triad. These collect bile and empty into the *common hepatic duct*. Excess bile enters the gallbladder through the *cystic duct*. The *gallbladder* lies in a pocket to the right of the quadrate lobe. It stores bile and contracts and squirts bile through the cystic duct into the *common bile duct*. The common bile duct carries bile from the gallbladder and liver to the duodenum where it enters through the *sphincter of Oddi*.

> Pylorus of stomach empties into first part of small intestine (duodenum)
> Secretions from liver, gallbladder, and pancreas enter at sphincter of Oddi
> Common hepatic duct joins gallbladder cystic duct in common bile duct

When needed, bile is released from the liver and squirted out from the gallbladder as it contracts. Bile from the gallbladder lows through the cystic duct to join the hepatic bile ducts in the common bile duct. Ducts from the pancreas, liver, and gallbladder enter the duodenum at the sphincter of Oddi. After bile has completed its role in triglyceride digestion, bile salts are recycled back to the liver through the hepatic portal vein (the *enterohepatic circulation*).

Hepatic Lobules

Describe the structure of hepatic lobules

The *hepatic lobule* is the functional unit of the liver. It consists of *hepatocytes* (liver cells) around a central vein. At the corners of the lobules are *hepatic triads*: a bile duct and branches of the *hepatic portal vein* and *hepatic artery*. The hepatic portal vein carries nutrients and old red cells to the liver for processing by its hepatocytes. It is a portal vein because it connects two capillary beds: one in the small intestine and the second in the liver. Hepatic arteries carry oxygenated blood to the liver. They join portal veins to form sinusoids that empty into the central vein in the center of the hepatic lobule.

> Liver consists of hepatic lobules arranged around central veins
> Portal veins bring nutrient-rich blood from small intestine
> Bile ducts carry bile from liver into hepatic bile duct

Hepatic sinusoids have a fenestrated endothelium with large, porous holes and no basement membrane. The *space of Disse* is between the hepatocytes and sinusoids. Here, blood directly bathes hepatocytes with plasma. There is no interstitial space between the sinusoid and the liver cells. The *central vein* collects blood from the sinusoids to return it to the circulation (back to the heart). *Kupffer cells* line the sinusoids. These phagocytic cells engulf pathogens and damaged cells.

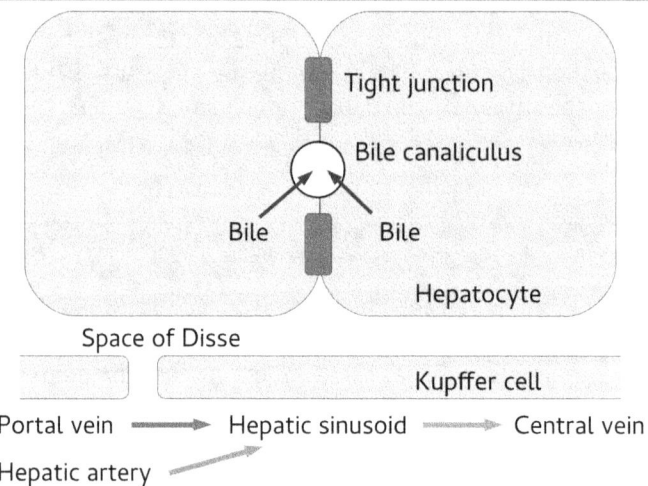

Nutrients from small intestine pass through sinusoids that
expose hepatocytes directly to plasma in space of Disse
Portal blood mixes with arterial blood to empty into central vein

Bile is secreted into canaliculi between hepatocytes that are joined together by tight junctions. *Bile* contains bile salts, bilirubin, lecithin
(phosphatidylcholine), and cholesterol. *Bile acids* are synthesized from
cholesterol in hepatocytes and conjugated with taurine or glycine to
make bile salts. *Bilirubin* is a yellow breakdown product of the hemoglobin pigment (heme) from old red cells that is produced by the
spleen.

Pancreas

Describe the secretions of the pancreas

The *pancreas* produces *bicarbonate* to neutralize the acidic chyme and
digestive enzymes that require a neutral pH to be active. These pass
through the pancreatic duct to enter the duodenum. The pancreatic duct
joins common bile duct which carries bile and pancreatic secretions.
The sphincter of Oddi controls release of these secretions into the duodenum.

Pancreatic acinar cells secrete digestive enzymes into ductules
lined with bicarbonate-secreting cells that merge into pancreatic duct
Scattered islets of Langerhans secrete glucagon and insulin into blood

Acinar cells of the pancreas secrete digestive enzymes (the *exocrine
pancreas*). Cells lining the ducts secrete bicarbonate. this bicarbonate
neutralizes the acidic chyme coming from the stomach. The pH opti-
mum of *pancreatic enzymes* is slightly basic. These enzymes include

proteases, lipases, and amylases. *Pancreatic proteases* (trypsin, chymotrypsin, carboxypeptidase) continue protein digestion. *Pancreatic lipase* breaks down triglycerides; it requires bile salts from the liver and *colipase*, which is secreted by the pancreas. *Pancreatic amylase* breaks down glycogen and continues starch digestion to disaccharides.

Small Intestine

Describe the overall structure and functions of the small intestine

The small intestine is divided into three regions: duodenum, jejunum, ileum. Secretions of the liver, gallbladder, and pancreas enter the duodenum to neutralize chyme and continue the process of digestion. The jejunum and ileum also have intestinal glands that secrete digestive enzymes, which are also distributed on cell surfaces. In the later parts of the small intestine, nutrients are absorbed.

> Small intestine is divided into duodenum, jejunum, and ileum
> Duodenum receives secretions of accessory organs
> Digestion continues and absorption occurs in jejunum and ileum

Throughout the small intestine, chyme is mixed and propelled by segmentation and peristalsis. *Segmentation* involves contraction of circular muscle at random spots in the small intestine to mix its contents. Materials pass through the small intestine by peristalsis, a moving wave of contraction identical to that in swallowing. Peristalsis occurs over short distances in the small intestine, so chyme has sufficient time for mixing with enzymes, digestion, and absorption.

> Peristalsis causes movement of intestinal contents along the tube
> Segmentation are random contractions that ensure mixing of contents

Intestinal Villi

Describe the structure of an intestinal villus and the brush border

The intestinal wall is highly folded. The first level of folding are the *plica circulares*. These intestinal folds are coated with *intestinal villi* that contain the specialized cells of the small intestine. Absorptive cells (*enterocytes*) have *microvilli* that form a *brush border* that further increases the surface area for absorption of nutrients.

> Intestinal wall has large folds (plica circulares) further folded into villi
> Villi are covered with intestinal epithelium with capillary and lacteal
> Secretory cells of intestinal crypts produce digestive enzymes

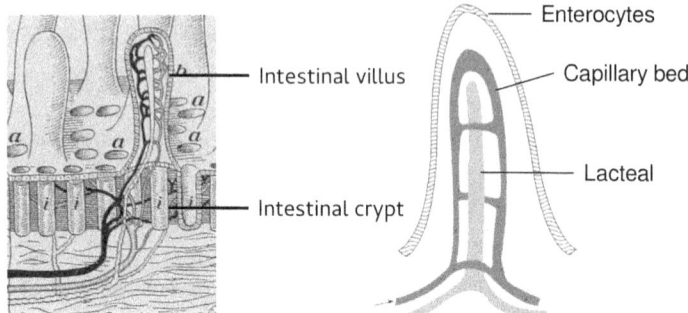

The apical surface of an enterocyte faces the *intestinal lumen*. Its basolateral surface is in contact with extracellular fluids. Cell junctions between the cells form a tight barrier between the apical and basolateral surfaces. In addition to absorptive cells, goblet (mucous) cells are present in the villi. These secrete protective mucus. Inside the villus is an arteriole, venule, and *lacteal* (lymphatic vessel) into which absorbed nutrients enter to be distributed throughout the body.

> Intestinal epithelial cells have microvilli that form a brush border
> Tight junctions seal cells together near their apical surface and
> separate apical from basolateral transporters and channels

Colon

Describe the role of the large intestine and its movements

Contents of small intestine pass from the ileum through the *ileocecal valve* to enter the *cecum* of the large intestine (*colon*), which is subdivided into ascending, transverse, descending, and sigmoid colon plus the rectum and anal canal. Longitudinal ribbons of muscle (*taenia coli*) that are always contracted to form permanent bulges (haustra). *Haustra* contract to propel the contents of the colon from cecum to rectum.

> Contents of small intestine enter large intestine through ileocecal valve
> Taenia coli are longitudinal ribbons of muscle that contract lengthwise to
> produce haustra (bulges); chyme is squeezed from one haustra to the next

Intestinal absorptive cells of the colon are in deep pits (*colonic crypts*). Colonic crypts secrete a lubricating mucus. The large intestine completes absorption of water (by osmosis) to produce a relatively dry feces. Colonic bacteria (*gut flora*) partially degrade foods that are not digested by pancreatic enzymes (fiber). This fermentation of fiber produces short chain fatty acids that are absorbed and enter capillaries. Vitamin K1 is provided by green leafy vegetables in the diet. Gut flora convert it to long-chain isoforms of *vitamin K2*, calcium-binding proteins required for synthesis of clotting factors.

> Colonic crypts of large intestine are deep pits lined with
> absorptive cells and goblet cells that secrete mucus
> Gut flora (bacteria) digest fiber and produce vitamin K

Defecation

Describe the defecation reflex

The large intestine undergoes *mass peristalsis* that pushes colonic contents from cecum to rectum. These waves of peristalsis move feces into the rectum some 3-4 times per day. They are distinct from the short peristaltic movements characteristic of the small intestine. *Feces* (indigestible materials) are temporarily stored in the rectum before excretion through the anus.

> Mass peristalsis waves (3-4 per day) push feces from colon into rectum
> Exit through anal canal protected by internal and external anal sphincters

Defecation is controlled by two sphincters. The *internal anal sphincter* relaxes in response to pressure in a filled rectum (parasympathetic defecation reflex). Stretch receptors send signals to the sacral spinal cord where the reflex is integrated. Motor neurons activate peristaltic waves in the colon and relax the internal anal sphincter. The *external anal sphincter* is normally under voluntary control. If the time is not appropriate for rectal evacuation, the contents of the rectum will return to the lower colon by *reverse peristalsis* for further absorption of water. Normally, the rectum is empty.

> Distension of rectum activates stretch receptors causing parasympathetic
> spinal defecation reflex that increases colonic peristalsis
> Internal sphincter opens; external sphincter is under voluntary control

Review Questions

1. What hormones control appetite?
2. What are the layers of the GI tract?
3. What is the difference (anatomical and functional) between myenteric and submucosal plexuses?
4. What is the enteric nervous system?

5. How do long reflexes differ from short reflexes? Where are short reflexes commonly observed?

6. What are the functions of saliva?

7. What happens during swallowing?

8. Why is gastric mixing important and what is partially digested food that leaves the stomach called?

9. How and why are only small amounts of chyme released into the duodenum?

10. What do each of the three cell types of gastric glands secrete?

11. What hormones control gastric secretions and motility?

12. How is the stomach lining protected from the damaging effects of gastric acid?

13. What are the three phases of control of gastric activity?

14. What do each of the cells in the pancreas secrete?

15. What is secreted into the duodenum?

16. What is the composition of bile?

17. What structure carries nutrients from the small intestine to the liver?

18. What is segmentation and what role does it play in intestinal digestion?

19. What is the defecation reflex?

Digestion and Absorption

In this lecture, we look into the details of how macronutrients are digested and the mechanism by which the resulting products of digestion are absorbed by the enterocytes that line intestinal villi. As we have seen before, the details require a molecular understanding of the process. We discover the *enzymes*, *transporters*, and *channels* that control the digestive and absorptive processes, their location and functions. After describing the mechanisms of absorption, we will consider how the body handles the nutrients that are absorbed (*metabolism*) in the next lecture.

Before we consider these molecular details, we look into how *appetite* is controlled by hormones produced by the stomach and adipose tissues. Our understanding of appetite is still rather meager. The mechanisms are rather complex. The goal is to give you a feel for how appetite is controlled without an overly elaborate explanation. In the end, weight is maintained when *calories* in food is equal to calories expended by the body in metabolism.

Calories

Define calories and how many calories different foods provide

Metabolism is the sum of all the reactions that occur in an organism. Metabolism distributes nutrients to pathways that provide and store energy or renew and build tissues. Metabolism is inefficient and always produces heat as a by-product.

The energy of foods is measured in *Calories* (big calories = kcal), the heat produced when a food is burned. Energy input and output in calories must balance to maintain constant weight. Carbohydrates and proteins provide 4 Cal/g while fats are more calorie-dense, providing 9 Cal/g. The lower energy yield of carbohydrates and proteins is largely due to the water that is associated with them. Fats are a more efficient way to store calories since they repel water and are packed into droplets.

> Calories measure the energy content of foods
> Proteins and carbohydrates produce 4 Cal/g and fats produce 9 Cal/g
> Calorie input must match calorie use to maintain body weight

Appetite

Describe the hormones that influence food intake

Appetite is our desire for food. The digestive system extracts as much nutrition as possible from meals. This means that our intake of calories is regulated by our appetite and not by the digestive system. Appetite is controlled by hunger and *satiety centers* in the hypothalamus. It is influenced by hormones produced by the GI tract and adipose tissues. The cerebral cortex and limbic system have additional input to appetite centers. The sight, smell, and taste of food increases gastric and salivary secretions. These gastric secretions prepare the stomach for incoming food.

> Ghrelin released by stomach acts on hypothalamus to cause hunger
> Leptin released by adipose tissue reduces hunger

Stomach hormones (*ghrelin*) are secreted during fasting to increase appetite. *Leptin* (produced by adipocytes) and GI hormones are secreted in the fed state to decrease appetite. Appetite hormones modulate the release of factors that lead to feelings of hunger or satiety. *Orexigenic* molecules stimulate appetite and decrease in the fed state. *Anorexigenic* molecules suppress appetite and increase in the fed state.

Macronutrients

List the macronutrients required to maintain human health

Macronutrients in the diet provide building blocks for synthesis and sources of metabolic energy. *Essential nutrients* are those that must be provided in the diet because the body cannot make them. *Proteins* are the source of essential amino acids and nitrogen. Fats (*triglycerides*) provide essential fatty acids. Along with *carbohydrates*, fats are primary sources of metabolic energy.

> Macronutrients include proteins, which provide amino acids and nitrogen
> and carbohydrates and fats that provide energy; all are digested
> to circulating nutrients: amino acids, glucose, and fatty acids

The *minerals* calcium, phosphorus, potassium, sulfur, sodium, chlorine, and magnesium are also needed in relatively large quantities to support growth and metabolism. *Micronutrients* are required in only small quantities and include the vitamins, minerals, and possibly some phytochemicals from plants. Our focus in this lecture will be on the digestion and absorption of macronutrients.

Digestion

Explain the chemical mechanisms that digest macronutrients

Dietary macromolecules must be hydrolyzed to produce components that can be absorbed. This is achieved by mechanical and chemical events that go on together in each compartment of the digestive tract. For example, digestion begins in the mouth with chewing of food (mastication) that breaks it down into particles that can be hydrolyzed by digestive enzymes. At the same time, dietary starches begin to be digested by salivary amylase.

> Digestion of carbohydrates begins in the mouth with amylase
> Protein digestion begins in the stomach with pepsin
> Fat digestion requires hepatic bile to begin digestion in duodenum

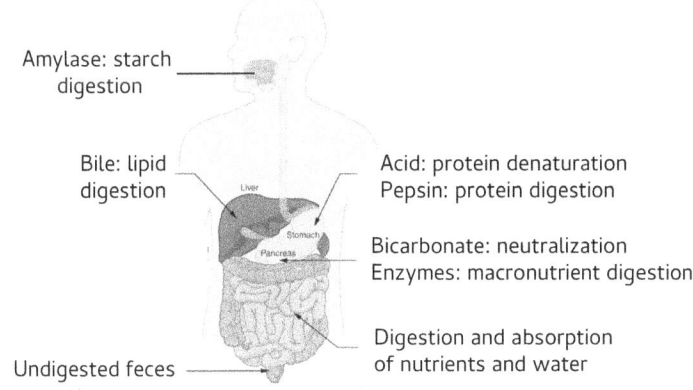

Digestive enzymes add water to the bonds that hold together the simple subunits of macronutrients. This hydrolysis breaks the bonds, producing simpler molecules that are either directly absorbed or further hydrolyzed by other enzymes.

> Digestion breaks down large molecules into smaller molecules by
> adding water across chemical bonds (hydrolysis) using -ases

Hydrolytic enzymes are named for their substrates and have the suffix *-ase*. Thus, *proteases* and peptidases hydrolyze proteins into absorbable peptides and amino acids. *Lipases* hydrolyze triglycerides into fatty acids and monoglycerides. *Amylases* hydrolyze starches and glycogen into disaccharides that are split into simple sugars by specific disaccharidases (sucrase, lactase, and maltase) located on intestinal epithelial cells. Nucleases hydrolyze nucleic acids into nucleotides. All of these enzymes are secreted into the duodenum to begin the intestinal stage of digestion.

Gastric Acid Secretion

Describe the mechanism of gastric acid secretion

A *proton pump* (H-K ATPase) pumps acid into the stomach lumen in exchange for K ions. The stomach lining is protected from damage by gastric acid by a bicarbonate-rich mucous layer (*mucosal barrier*) that lines the gastric mucosa and is secreted by gastric mucous cells. Pumping of acid into the stomach is balanced by secretion of bicarbonate from the stomach into the blood. This alkaline tide causes an increase in the pH of the blood leaving the stomach.

> H ions and bicarbonate are produced by intracellular carbonic acid
> Apical H-K ATPase of parietal cells pumps H ions into stomach lumen
> Bicarbonate is secreted into interstitial fluids (alkaline tide)

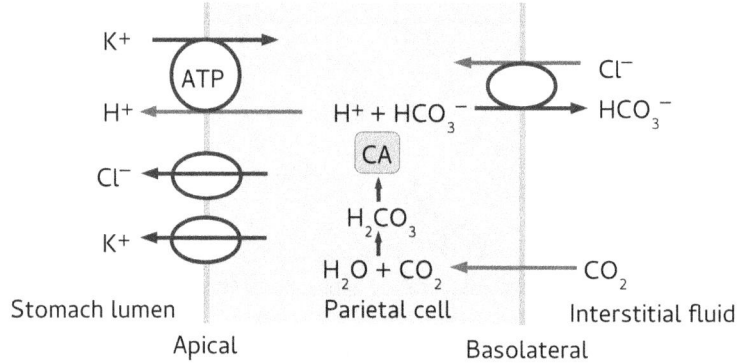

> Gastric mucous neck cells secrete mucus and bicarbonate
> to protect stomach lining from harsh acidity of gastric acid
> Somatostatin released by D cells of antrum inhibit gastric acid secretion

Gastrin is released in response to distension (via stretch receptors) and the presence of foods (via chemoreceptors). Gastrin enters the bloodstream to stimulate secretion of gastric acid by G cells that are primarily located in the antrum. In addition, The parasympathetic vagus nerve (CN X) secretes acetylcholine (ACh) which stimulates acid secretion directly from parietal cells and indirectly by stimulating the ECL cells to release histamine. *Histamine* released by the ECL cells stimulates gastric acid secretion from parietal cells. Histamine binds to the H2 receptor on parietal cells. This receptor is distinct from the H1 receptor responsible for allergic rhinitis (runny nose and sneezing).

Peptic ulcers are open sores in the mucosal lining of the gastrointestinal tract. They may be gastric or more commonly duodenal and are due to damage by stomach acids. Gastric bleeding can be life-threatening. *Helicobacter pylori* (H. pylori) infection is a common cause. This gram-negative bacterium infects stomach mucosa. It survives the high acidity of the stomach lining, then destroys the gastric endothelial barrier and causes inflammation.

G cells in antrum of stomach release gastrin that activates ECL cells
Enteric nervous system directly activates acid secretion and
ECL secretion of histamine that activates apical H-K pump

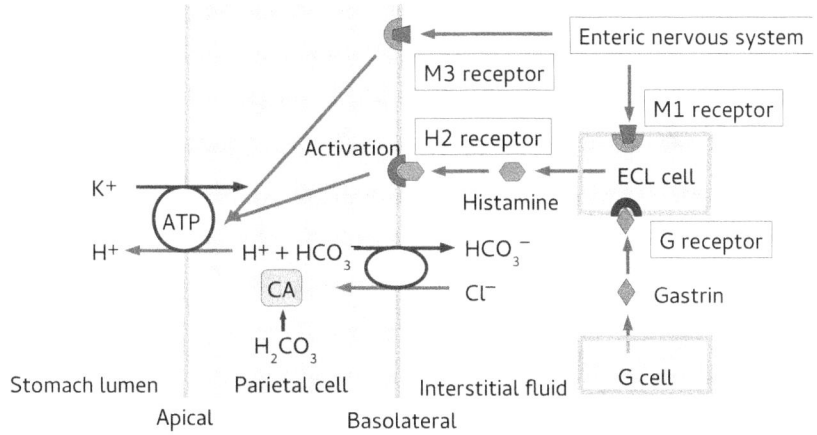

GERD (*gastroesophageal reflux disease*) is an esophageal mucosal irritation due to reflux of gastric acid through a weak cardiac sphincter. It is commonly called *heartburn. Erosive esophagitis* in an inflamed and ulcerated esophagus which is usually due to acid reflux.

Acid secretion can be reduced to minimize progression of peptic ulcers and reduce the symptoms of gastroesophageal reflux disease (GERD) by either inhibiting the proton pump directly (*omeprazole*) or blocking the histamine H2 receptor with a specific antagonist (*ranitidine*). Antibiotics (*tetracycline* or *amoxicillin*) are added to acid inhibitors when H. pylori infection is present.

Prostaglandins (PGE2) decrease gastric acid secretion and increase mucus and bicarbonate production. PGE2 is produced by oxidation of arachidonic acid. Protective prostaglandins are produced by COX-1 (*cyclooxygenase*-1) and act locally as paracrines. NSAIDs (nonsteroidal anti-inflammatory drugs, such as *aspirin* and *ibuprofen*) nonspecifically inhibit cyclooxygenase enzymes. Inhibition of COX-1 by NSAIDs decreases prostaglandin production and stimulates gastric acid secretion and minimize its protective coating. Inflammation increases COX-2 production of prostaglandins. COX-2 selective anti-inflammatories (*acetominophen*) minimize inflammatory prostaglandins that cause fever and pain without damaging the gastric mucosa.

Intrinsic factor is a glycoprotein secreted by parietal cells of the stomach. It binds to vitamin B12 so that it can be absorbed in the small intestine.

> Gastric parietal cells secrete gastric acid and intrinsic factor
> Intrinsic factor secretion is required for intestinal vitamin B12 absorption

Gastric Protein Digestion

Describe how protein digestion begins in the stomach

Gastric acid causes proteins to unfold (denature), exposing their peptide bonds to proteases. Pepsinogen is activated by *pepsin* at low pH and begins digestion of proteins in the stomach. Gastric mixing is required to mix stomach contents with gastric acid to ensure proteins are fully denatured. The partially digested proteins in the stomach make it more liquid so that small volumes of chyme can be released through the pyloric sphincter into the duodenum.

> Pepsinogen is activated to pepsin in low pH of stomach
> Gastric acid unfolds proteins (denaturation) exposing peptide bonds
> Protease activity of pepsin digests proteins into smaller peptides

Digestive Hormones

Describe the hormones that control pancreatic and hepatic secretions

The low pH of chyme and the presence of lipids and carbohydrates in the duodenum stimulate secretion of several hormones that activate secretions and inhibit gastric activity. Decreased pH in the duodenum, as a result of the incoming acidic chyme, stimulates secretin release. *Secretin* inhibits secretion of gastric acid and stimulates pancreatic bicarbonate secretion. Pancreatic bicarbonate neutralizes incoming acidic chyme so that intestinal digestive enzymes can function effectively.

> Chyme slowly entering duodenum triggers secretion of hormones
> In response to low pH secretin increases secretion of bile from liver and
> bicarbonate from pancreas that provides optimal pH for intestinal enzymes

CCK (*cholecystokinin*) and GIP/GLP-1 are released in response to nutrients in the duodenum. CCK stimulates the pancreas to produce digestive enzymes, the liver to produce bile, and the gallbladder to secrete bile. GIP (*gastric inhibitory peptide*) inhibits gastric mobility. Motilin increases motility of the small intestine.

Cholecystokinin (CCK) released in response to lipids and carbohydrates in duodenum passes through blood to activate secretion of bile from liver and gallbladder and pancreatic enzymes and inhibit gastric activity

The *incretins*, GIP and GLP-1 stimulate insulin release and reduce glucagon secretion from the pancreas in anticipation of a surge in blood glucose. Endothelial dipeptidase (DPP-4) cleaves a dipeptide from incretins, inactivating them. Inhibitors of DPP-4 (*sitagliptin*) maintain elevated incretin levels which decreases blood glucose levels.

Incretins (GLP-1 and GIP) released into blood in response to food increase insulin in anticipation of sugar surge
Endothelial DPP-4 inactivates incretins (inhibitors lower blood glucose)

Carbohydrate Digestion

Describe the steps of intestinal digestion of carbohydrates

In the duodenum, pancreatic enzymes continue the digestion of macronutrients. Carbohydrates are digested to disaccharides and proteins to small peptides. Triglycerides are digested to monoglycerides and fatty acids. After digestion and absorption, the remaining undigested materials enter the cecum of the large intestine through the ileocecal valve.

Carbohydrate digestion begins with *salivary amylase* which breaks down starches to sucrose, so that chewing a starchy food makes it taste sweeter. *Pancreatic amylase* secreted into the duodenum along with bicarbonate continues digestion of carbohydrates to disaccharides (two simple sugars linked together).

Salivary amylase begins digestion of complex carbohydrates (starches) Pancreatic amylase completes digestion to disaccharides that are hydrolyzed by specific disaccharidases on intestinal epithelium

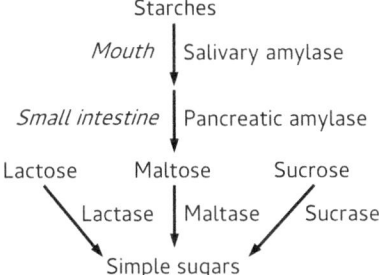

Intestinal glands in the wall of the small intestine also produce digestive enzymes. Brush border enzymes break down disaccharides (and small peptides). *Maltase* digest the disaccharide maltose to glucose;

sucrase digests sucrose to glucose and fructose; *lactase* digests lactose to glucose and galactose. The presence of lactase in adults is genetically determined. Its absence results in lactose intolerance, the inability to digest foods (such as milk products) that contain lactose.

Carbohydrate Absorption

Describe the steps of intestinal absorption of carbohydrates

Simple sugars are taken up into cells through apical transporters. Fructose enters directly through a GLUT transporter. Glucose and galactase enter with Na ions through the SGLT (*sodium-glucose linked transporter*) which is driven by the Na gradient established by a basolateral Na pump (secondary active transport).

> Simple carbohydrates are absorbed by apical transporters (luminal side)
> Glucose and galactose require Na pump for secondary active transport
> Fructose enters through transporter; all exit by facilitated diffusion

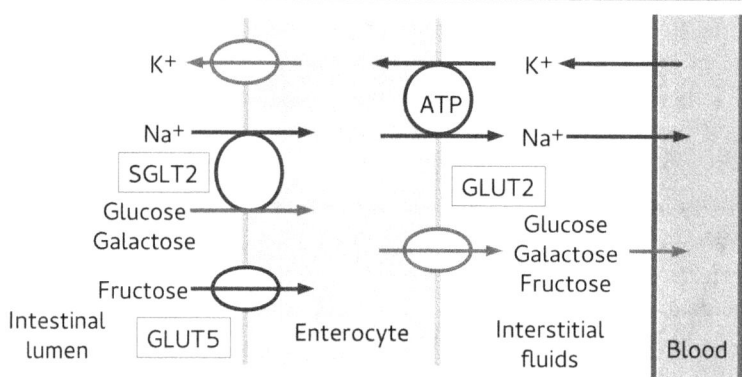

From the enterocyte, simple sugars pass into interstitial fluids through a *GLUT2 transporter*. From interstitial fluids, sugars (and amino acids) enter capillary beds. Blood vessels from the intestinal villi pass directly to the liver through the hepatic portal vein.

Protein Digestion and Absorption

Describe the process of protein digestion and absorption

Protein digestion begins in the stomach with unfolding of protein chains by *gastric acid*. Some of the exposed peptide bonds are cleaved by *pepsin*. This produces a partially digested protein. A set of *pancreatic proteases* released into the duodenum continue digestion to small peptides and amino acids at neutral pH. Peptides with two or three amino acids can be directly absorbed into intestinal cells (enterocytes). through H ion-linked transporters. Peptidases on the brush border of enterocytes and intracellular peptidases digest small peptides to amino acids. Only amino acids can be used by the body.

> Proteins in stomach are denatured by acid and hydrolyzed by pepsin
> Pancreatic proteolytic enzymes continue digestion to small peptides
> which are fully digested to amino acids by intestinal enzymes

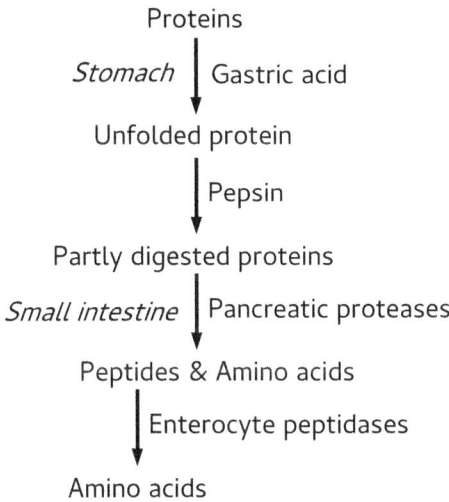

Amino acids enter enterocytes through a set of apical transporters linked to Na ion gradients (secondary active transport). Several baso-lateral transporters allow amino acids in enterocytes to pass into interstitial fluids and then into capillaries which lead to the liver through the hepatic portal vein.

> Amino acids and short peptides enter enterocytes by secondary active
> transport; short peptides are degraded to amino acids
> and enter interstitial fluids through facilitated diffusion

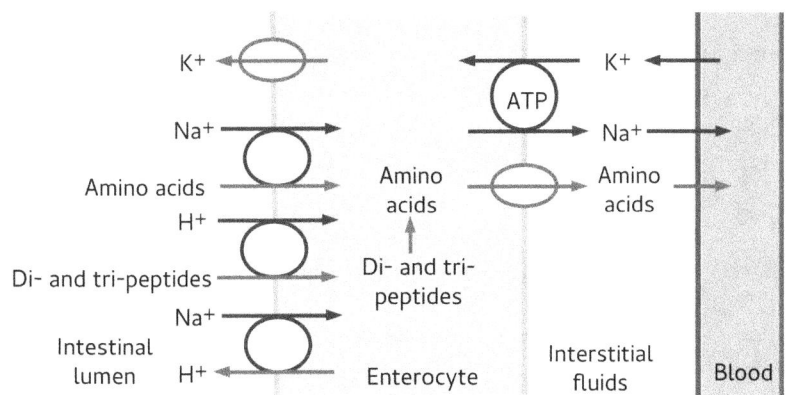

Triglyceride Digestion

Describe the role of bile salts in triglyceride digestion

Triglycerides are not soluble in water. Chewing fats begins to mechanically break them up into smaller droplets, a process which is continued in the stomach. However, these droplets are not easily digested. They must first be made into very tiny particles (*micelles*) by combining with bile salts in hepatic bile. This *emulsification* allows pancreatic lipase to digest the *triglycerides* (glycerol with three fatty acids) to two fatty acids and a monoglyceride (glycerol with one fatty acid). Pancreatic lipase also requires a small protein (colipase), that is secreted by the pancreas, to be fully active. Fatty acids and monoglycerides are directly absorbed into enterocytes by passing through the cell membrane.

> Lipids (triglycerides) are insoluble and must be emulsified
> by bile acids to be digested by pancreatic lipase
> Lipase cleaves triglycerides to fatty acids and monoglycerides

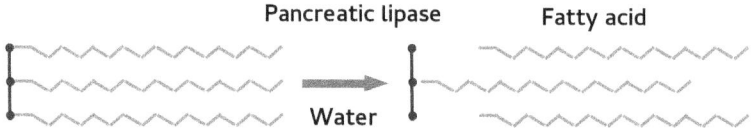

Intestinal Lipid Transport

Explain how the products of triglyceride digestion enter the circulation

Water-soluble nutrients (amino acids, simple sugars, and short-chain fatty acids) enter the capillaries of intestinal villi and pass through the hepatic portal vein to the liver. Fatty acids that are absorbed by enterocytes follow a more complex pathway. The key to understanding the metabolism of triglycerides is to understand that the body cycles between triglycerides and fatty acids. Triglycerides are a storage form found in adipocytes (fat cells) and the interior of small particles in the blood called lipoproteins. Fatty acids must be released from triglycerides to enter and be used by cells to produce energy.

Fats in the form of fatty acids enter intestinal absorptive cells. There fatty acids are synthesized into triglycerides, the storage and transport form. Since triglycerides are not soluble in water (they form droplets), they are packaged into protein and phospholipid coated chylomicrons, a type of lipoprotein (which combines lipids with proteins). Chylomicrons are large particles and cannot enter capillaries. Rather, they are secreted from enterocytes by exocytosis and enter the lacteals, forming a milky chyle. The lacteals merge to form larger lymphatic vessels that empty into the left subclavian vein through the thoracic duct.

> Fatty acids and monoglycerides enter intestinal cells where they are converted to triglycerides and assembled with proteins and cholesterol into chylomicrons that leave cells by exocytosis to enter lacteals

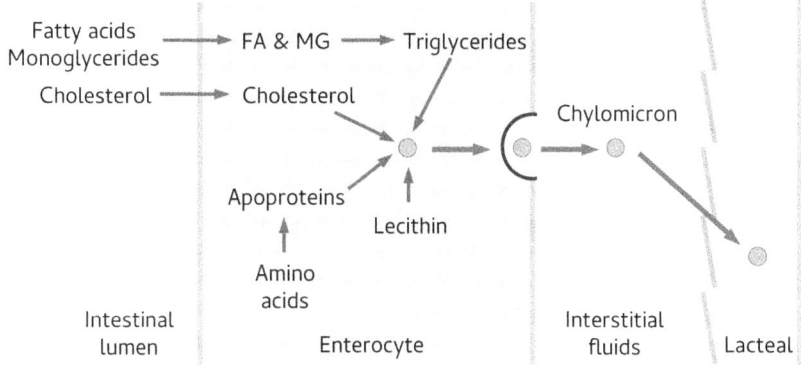

Intestinal Absorption

Explain how products of digestion are transported

Water-soluble nutrients (simple sugars and amino acids) are absorbed into capillary beds of intestinal villi. They are transported through the hepatic portal vein directly to the liver, which processes the nutrients, depending on overall body needs. There, most of the fructose and galactose are converted to a glucose intermediates (fructose-1-P and glucose-6-P). Chylomicrons are too large to enter capillaries, so enter lacteals of the intestinal villi, where they pass through lymphatic vessels to the left subclavian vein.

> Water-soluble nutrients (simple sugars, amino acids, short fatty acids) enter capillaries and are transported to liver by hepatic portal vein
> Chylomicrons enter lacteals to be transported to left subclavian vein

Water Absorption

Describe how water is absorbed in the intestine

Most of the water taken in by ingestion and from digestive secretions is absorbed in the small intestine. Additional water is absorbed in the large intestine. *Water absorption* is driven by osmosis because of the Na gradient established by the *basolateral Na pump*. Na pumped out of cells establishes an osmotic gradient wherein interstitial fluids are hyperosmotic relative to the fluids within the intestinal lumen. Thus, water follows the osmotic gradient from lumen across enterocytes, into the interstitial fluids and then into capillaries.

> Water absorption is driven by secondary active transport of Na ions across intestinal epithelial cells that establishes osmotic gradient
> Most water is reabsorbed in small intestine and the remainder in colon

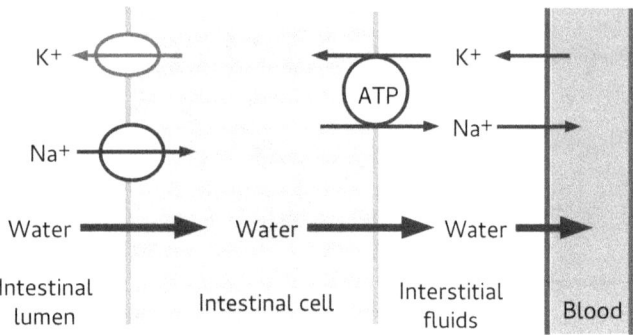

Review Questions

1. What are the three types of macronutrients and in what forms are they absorbed in the intestine?

2. Where and how does digestion begin for carbohydrates, proteins, and lipids?

3. What is the primary chemical mechanism for digestion?

4. Why is a low gastric pH required to begin digestion of proteins?

5. Why are pancreatic bicarbonate secretions important?

6. Describe the stages of carbohydrate digestion from mouth through intestine.

7. What enzymes of the intestinal brush border are responsible for digestion of disaccharides to simple sugars?

8. How are simple carbohydrates absorbed from the intestinal lumen into interstitial fluids?

9. Describe the stages of protein digestion from stomach through intestine.

10. What is required for fat digestion?

11. What happens to absorbed fatty acids? In what form are they transported into lacteals?

12. How does transport of water-soluble nutrients and chylomicrons differ?

13. How and where is water absorbed in the intestines?

Whole-Body Metabolism

Most of our discussion of cellular metabolism has focused on how *ATP* is provided to muscle cells for contraction. As you may recall, resting skeletal muscle fibers, smooth muscle cells, and cardiomyocytes use a nearly equal supply of *fatty acids* and *glucose* to produce ATP. Glucose is metabolized to *pyruvate* by *glycolysis*. Pyruvate enters mitochondria to produce the maximum amount of energy possible in the presence of adequate oxygen (*aerobic metabolism*). Fatty acids are always oxidized in mitochondria (*beta-oxidation*). When skeletal muscles are contracting with maximum force, they convert glucose to lactate by *anaerobic metabolism*.

This quick summary tells us what happens in the cells. But many questions remain. How are glucose and fatty acids supplied to the cells? What happens to the lactate that is produced during anaerobic metabolism? What is the source of amino acids for protein synthesis? How are glucose levels in the blood regulated?

These are questions of whole-body metabolism. They require an understanding of how metabolic reactions differ in different tissues. The *liver* plays a central role in whole-body metabolism. It is one of the few tissues able to use dietary fructose and galactose. It ensures that *blood glucose* levels do not fall too low by storing glucose as *glycogen* and releasing glucose into the blood when needed. The liver converts *lactate* and other metabolites to glucose. It provides *ketone bodies* for brain metabolism when glycogen stores are depleted in starvation.

Insulin ensures that skeletal muscles and adipose tissues get adequate glucose from the blood. Glycogen is stored in *skeletal muscles* to provide glucose intermediates that are metabolized for energy, as needed. *Adipose tissues* store fatty acids as triglycerides. When the body requires fatty acids, they are released from these triglyceride stores. Triglycerides are also provided to tissues by dietary chylomicrons and hepatic VLDL. These *lipoproteins* form part of a complex set of interactions in the blood that carry cholesterol between tissues and the liver.

Finally, I discuss the roles of glucagon in maintaining adequate blood glucose levels, epinephrine and glucocorticoids in the stress response, and the response to long-term food deprivation.

Metabolic Interconversions

Summarize the metabolic pathways that modify dietary nutrients

Our diets are not ideal. The body must use what it gets, so many of the nutrients we ingest must be converted to more suitable molecules. The primary simple sugar used by the body is *glucose*. In lactating mothers, dietary *galactose* is incorporated into milk sugar (lactose) in the mammary glands. Sperm use fructose for energy production. Galactose is also a common sugar component of glycoproteins and glycolipids. Otherwise, galactose and fructose are metabolized primarily by the liver during first-pass metabolism (directly after intestinal absorption). Galactose enters the metabolic pathway for glucose and is subject to the same close hormonal control.

Fructose is rapidly metabolized by the liver, where it is taken up by the same GLUT2 transporter as glucose and galactose. Sorbitol (a sugar substitute) is converted to fructose in the liver. This conversion requires energy, so sorbitol has less energy content than ordinary sugar molecules.

Circulating levels of fructose are very low in comparison with glucose. In the liver, the products of fructose metabolism are glycogen and fatty acids. Fructose also provides glycerol for synthesis of triglycerides. When liver glycogen is plentiful, nearly all of fructose is directed toward triglyceride synthesis. High fructose diets elevate triglycerides in the blood (hypertriglyceridemia).

Humans can synthesize 11 of the 20 amino acids. Those that cannot be synthesized (phenylalanine, valine, threonine, tryptophan, methionine, leucine, isoleucine, lysine, and histidine) are *essential amino acids* in the diet. The nucleotides required to produce ATP, nucleotide signaling molecules, DNA, and RNA can all be synthesized. In all cases, the primary source of nitrogen required for synthesis of amino acids and nucleotides are dietary amino acids.

Glucogenic amino acids (all but leucine and lysine) can be used to produce glucose by gluconeogeneis (production of new glucose) in the liver. The brain uses glucose for energy and not fatty acids (they are not able to pass through the blood-brain barrier). *Ketogenic amino acids* can be converted to ketone bodies during starvation, providing the brain with an alternative source of energy.

Palmitc acid (a fatty acid with 16 carbons) can be synthesized directly in the cytoplasm from products of sugar metabolism. It can be elongated to produce stearic acid. However, double bonds cannot be introduced beyond carbon 9 or 10. For this reason, linolenic acid and linoleic acid, which have double bonds near the ends of their chains are *essential fatty acids*. Double bonds can be added to these essential fatty acids (EFAs) to produce polyunsaturated fatty acids, such as arachidonic acid, that are used to synthesize prostaglandins and leukotrienes which play important roles in inflammation.

The glycolytic pathway also provides glycerol for *triglyceride synthesis* (three fatty acids plus glycerol = triglyceride). It is not possible to synthesize glucose from fatty acids. Rather, the liver uses ATP produced by fatty acid oxidation to provide the energy required for new glucose synthesis (*gluconeogenesis*).

Nutrient Pools

Explain the concept of nutrient pool and list the circulating nutrients

Metabolism begins with the breakdown of dietary macromolecules into simpler molecules that can be absorbed: simple sugars, fatty acids, and amino acids. Simple sugars and amino acids are water-soluble and are transported in the plasma. Fatty acids bind to albumin in plasma and are incorporated into triglycerides that are carried by lipoproteins. *Lipoprotein lipase* on the endothelial cell surface of blood vessels hydrolyzes lipoprotein triglycerides to locally produce fatty acids. These simple molecules (glucose, amino acids, and fatty acids) make up a circulating *nutrient pool* that cells use for energy and as building blocks for more complex molecules.

> Nutrient pools are circulating nutrients: soluble glucose and amino acids plus fatty acids bound to albumin or in triglycerides of lipoproteins
> Nutrients enter cells for energy and synthesis of biomolecules

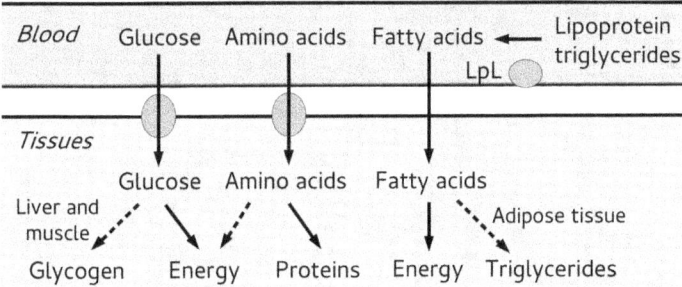

Water soluble nutrients (glucose and amino acids) pass through the endothelium lining capillaries through water-filled channels, largely between cells (paracellular transport). Once in the interstitial fluids, transporters mediate their entry into cells (facilitated transport). Fatty acids are lipid soluble. They can pass through membranes to cross the endothelial barrier and directly enter cells. This passive diffusion is complemented by fatty acid transporters in cell membranes. Once inside the cell, fatty acids bind to fatty acid binding proteins.

Glucose enters cells to produce energy and provide metabolic intermediates, or is stored in liver and skeletal muscle tissues as glycogen. *Amino acids* are used primarily for protein synthesis. In starvation, amino acids can be used to produce energy. Amino acid catabolism produces ammonia that is converted to *urea* in the liver, which is excreted by the kidneys. *Fatty acids* are used by skeletal and cardiac muscles for energy. Some are used to produce the phospholipids that make up cell membranes. Excess fatty acids are stored as *triglycerides* in adipose tissues.

Glycogen

Describe the role of glycogen in storage and release of glucose

Glycogen is the storage form of glucose in liver and skeletal muscle fibers. Glycogen breakdown (*glycogenolysis*) produces phosphorylated glucose intermediates. Skeletal muscle breaks down these intermediates in the glycolysis pathway to provide energy for cellular functions. Muscle tissues cannot convert glycogen to glucose.

Glucose is stored as glycogen in muscle and liver cells
Glycogen may be mobilized to enter glycolysis to produce energy

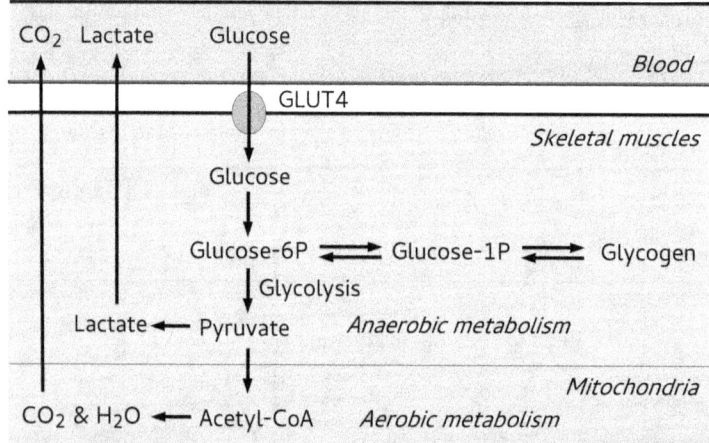

In the liver and kidney, the enzyme glucose-6-phosphatase is available to convert phosphorylated glucose molecules (glucose-6-phosphate) produced by glycogenolysis into free glucose. This glucose exits cells through the bidirectional GLUT2 transporter and and contribute to blood glucose. Glycogenolysis is stimulated in hepatocytes by glucagon in order to maintain blood glucose levels in-between meals.

Glycogenolysis produces glucose-6-phosphate and glucose in liver
Hepatic glucose enters blood to maintain blood glucose levels
Fructose produces glycogen and fatty acids in hepatocytes

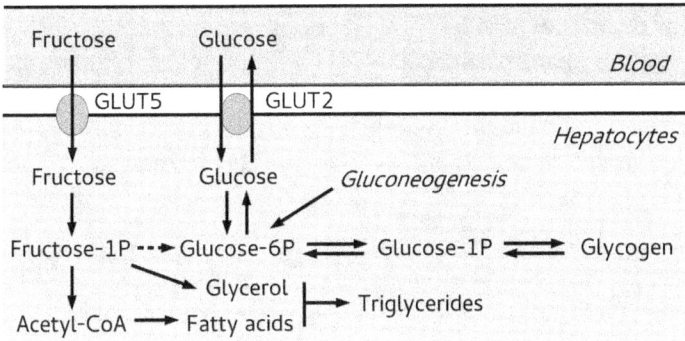

Triglycerides

Describe the storage and mobilization of fatty acids as triglycerides

In adipose tissues, fatty acids are stored as *triglyceride droplets*. The glycerol for synthesis comes from blood glucose. Fatty acids are supplied by hydrolysis of lipoprotein triglycerides by endothelial lipoprotein lipase. Once triglycerides are synthesized, they enter large droplets of fat that take up almost the entire cell.

Fatty acids are mobilized from adipose tissues by hormone-sensitive lipase. This intracellular enzyme is activated by epinephrine and glucagon. It hydrolyzes triglycerides to produce glycerol and fatty acids. The fatty acids leave the adipocytes and bind to serum albumin which transports them in the blood. In the liver, fatty acids are synthesized into triglycerides that are packaged for secretion in VLDL.

Fatty acids enter adipose tissue to be stored as triglyceride droplets
Fatty acids released into blood by hormone-sensitive lipase in
response to epinephrine or glucagon are carried bound to albumin

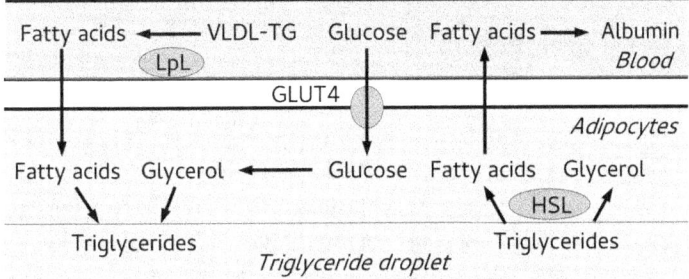

Lipoproteins

Define lipoproteins and explain their role in fat transport

Fatty acids are transported either bound to the abundant plasma protein albumin or as triglycerides. Triglycerides are insoluble, so are carried in the core of small particles covered with proteins and phospholipids (*lipoproteins*). Fatty acids are released from lipoproteins into the blood at tissues by the capillary endothelial enzyme *lipoprotein lipase*. Fatty acids are used by skeletal and cardiac muscles for energy. Adipose tissues store excess fatty acids as triglycerides.

> Fatty acids are carried from intestinal lacteals into blood by chylomicrons
> Lipoprotein triglycerides are hydrolyzed to fatty acids by endothelial LpL
> Hepatic triglycerides are carried by very low density lipoproteins (VLDL)

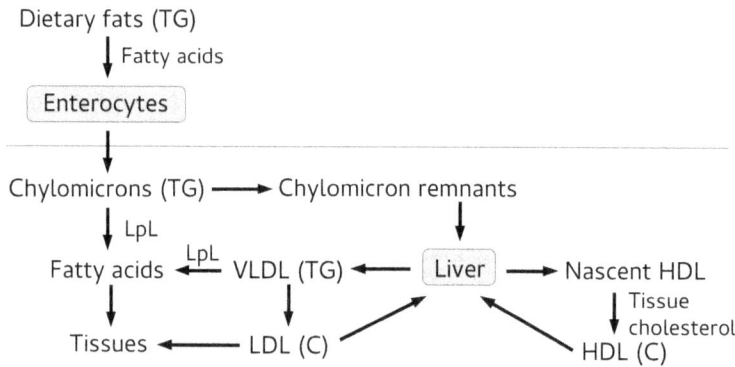

Lipoproteins have a core of triglycerides and cholesterol esters surrounded by a monolayer of phospholipids and proteins. *VLDL* (very low density lipoproteins) carry triglycerides synthesized in the liver. *Chylomicrons* carry dietary triglycerides. After hydrolysis of its triglycerides by lipoprotein lipase on the surface of capillary endothelial cells, chylomicrons become *remnants* that are taken up by the liver. The remnant of VLDL becomes a low density lipoprotein (LDL) that delivers cholesterol to cells.

Cholesterol Metabolism

Describe the role of lipoproteins in cholesterol metabolism

One of the remnants of VLDL is LDL (*low density lipoproteins*). HDL (*high density lipoproteins*) are produced in a nascent form by the liver. These are the major carriers of cholesterol in the body. Hepatic LDL receptors take up LDL by receptor-mediated endocytosis. These receptors are upregulated (increased in numbers) when hepatic cholesterol levels are low. Cholesterol is delivered to tissues for steroid and membrane synthesis primarily by LDL. HDL removes cholesterol from cells and transfers it to the liver.

High LDL cholesterol levels are associated with increased risk of coronary artery disease. Inhibition of cholesterol synthesis in hepatocytes by statins (*atorvastatin*), upregulates (increases expression of) LDL receptors in the liver. The increase in receptors clears LDL from the blood and reduces LDL cholesterol levels, resulting in decreased risk of lipid accumulation in arteries (atherosclerosis) and myocardial infarction.

Insulin

Describe the effect of blood glucose on insulin levels and how insulin decreases blood glucose

Blood glucose increases after meals. Increased blood sugar promotes insulin and GLP-1 release which stimulates glucose uptake and storage. Insulin is secreted by *pancreatic beta cells* in response to blood glucose levels. Beta cells are in the islets of Langerhans have GLUT2 receptors that allow glucose to pass into and out of the cells, so the intracellular concentration of glucose matches that in the blood. Intracellular glucose metabolism produces ATP that closes ATP-sensitive K ion channels, causing depolarization of the cell. The release mechanism for insulin is similar to that in axon terminals. Depolarization opens voltage-gated Ca channels causing Ca to enter cells and exocytosis of insulin from secretory vesicles.

GLP-1 (glucagon-like peptide 1) is produced in the intestine and brainstem in response to ingestion of food. This incretin increases insulin secretion, even in type 2 diabetics.

> Increased blood glucose after meal causes increased GLP-1 and
> insulin levels which reduce blood glucose levels back to normal
> Between meals, glucose is kept from falling too low by glucagon

Insulin is a small protein (a peptide). It binds to a cell membrane receptor that has a *tyrosine kinase* domain on its intracellular face. Tyrosine kinase receptors phosphorylate regulatory proteins that initiate a cascade of reactions within the cell, resulting in changes in enzyme activity and cellular function. They differ from GPCRs (G-protein coupled receptors) that are linked to small intracellular G proteins that activate or inhibit signaling cascades.

In adipose and muscle tissues, insulin causes insertion of GLUT4 transporters into the membrane. GLUT4 allows entry of glucose into cells by facilitated diffusion. This entry directly reduces blood glucose concentrations. Insulin also stimulates amino acid uptake and protein synthesis. In adipose tissues, insulin stimulates triglyceride deposition and inhibits lipolysis (triglyceride breakdown to fatty acids). This reduces fatty acid available for oxidation so that glucose becomes the preferred energy source in the body.

Insulin causes insertion of GLUT4 transporters in adipose and skeletal muscle tissues to increase entry of blood glucose

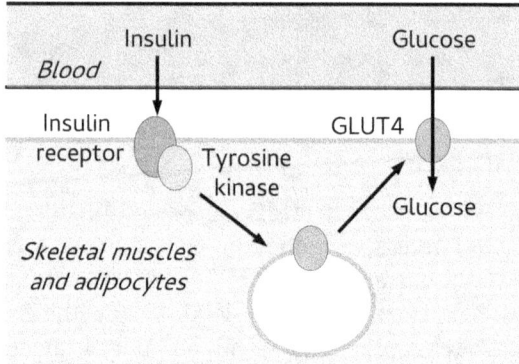

In the hepatocytes, glucose passes back and forth between cells and blood through GLUT2 transporters. When blood glucose is high, glucose enters hepatocytes. Insulin increases glucose phosphorylation, trapping it in the cells where it is used to synthesize glycogen (glycogenesis). At the same time, insulin inhibits gluconeogenesis. The net effect is storage of blood glucose in liver cells as glycogen.

Glucose enters hepatocytes through GLUT2 transporters
Insulin increases glucose phosphorylation, pulling glucose into the cell

Glucagon

Describe the effects of glucagon on blood sugars

Insulin returns elevated blood glucose to normal levels. *Glucagon* keeps glucose from falling too low during fasting. Glucagon stimulates release of glucose from the liver to increase blood glucose. In the *fed state*, nutrient pools are elevated (sufficient to provide energy needs). Fed state nutrients enter cells to be incorporated into triglycerides, glycogen, and proteins. With high blood sugar, insulin increases the use of glucose, ensuring it is stored for later use.

In the *fasting state*, nutrient levels are diminished. In fasting, storage forms of nutrients are catabolized to produce energy and maintain

blood glucose. With low blood sugar, glucagon mobilizes glucose reserves to maintain blood glucose.

> Gluconeogenesis produces glucose in liver from fatty acids,
> glucogenic amino acids, glycerol, or lactate

Glucagon acts on the liver to produce blood glucose by new synthesis
(*gluconeogenesis*). Glucagon causes glycogen breakdown (*glycogenol-ysis*) in the liver. Hepatic glycogenolysis and gluconeogenesis elevates
hepatic glucose in the fasting state. Elevated hepatic glucose enters
plasma by facilitated diffusion through the bidirectional GLUT2 transporters in the cell membrane. In addition, glucagon elevates lipolysis
to provide fatty acids as an alternative source of energy which preserves blood glucose for the brain.

> Glucagon mobilizes hepatic glycogen stores and
> increases gluconeogenesis to maintain blood glucose and
> increases fatty acid synthesis and lipolysis to elevate fatty acids

Diabetes Mellitus

Diabetes is a group of diseases that cause increased urine output. *Dia-betes insipidus* is due to a lack of antidiuretic hormone, which prevents
insertion of the aquaporin channels required to reabsorb water in the
distal tubules and collecting ducts of the kidney. *Diabetes mellitus*
means sweet urine. Blood glucose is elevated to such an extent that it
saturates the nephron glucose transporter (SGLT), preventing complete
glucose reabsorption. Glucose that is not reabsorbed is spilled into the
urine.

Diabetes mellitus may be due to low insulin levels (*type 1 diabetes*) or
lack of tissue responsiveness to insulin (*type 2 diabetes*). Type 1 dia-
betes can be treated with insulin injections, so has been called *insulin-dependent diabetes*. Its onset is commonly in childhood (so, *juvenile diabetes*). Usually type 1 diabetes is due to autoimmune destruction of
insulin-producing beta cells in the pancreas. Untreated, it causes *hy-perglycemia* (elevated blood glucose) and *diabetic ketoacidosis*
(overproduction of acidic ketone bodies).

Type 2 diabetes is not treated with insulin injections, so is also called
non-insulin dependent diabetes mellitus (NIDDM). Its onset is commonly in adults (*adult onset diabetes*) and is highly correlated with
obesity. In type 2 diabetes, target cells do not respond properly to insulin (*insulin resistance*). In response, beta cells increase insulin
production to make up for its low cellular activity. Treatment focuses
on making cells more responsive to insulin and interventions that re-

duce blood sugar levels. As the disease progresses, the beta cells wear out and are unable to produce enough insulin. Insulin is often needed as the disease progresses.

Screening tests for diabetes rely on *fasting blood glucose* levels. Normal fasting glucose should be < 110 mg/dL. Diabetes mellitus is suspected when fasting glucose is > 126 mg/dL. Between these values, the risk for developing type 2 diabetes is elevated (*prediabetes*). The *oral glucose tolerance test* (OGTT) is sometimes used as a follow-up to fasting glucose screening and in pregnant women. Blood glucose is measured 2 hr after a 75 g glucose challenge. Normal glucose should be < 200 mg/dL. Higher levels suggest diabetes mellitus.

Hemoglobin A1C measures glycosylation of hemoglobin due to long-term glucose elevation. This test is often performed when fasting glucose levels are elevated and to monitor glucose control in treated diabetics. The target value is < 6.5%. *Glycosylation* of hemoglobin is a marker for glycosylation of proteins in the microvasculature and glomerular membrane that disrupt their function and lead to the common symptoms of poorly controlled diabetes: cardiovascular and kidney disease, retinopathy, and neuropathic pain. These side effects are minimized if glucose levels are kept from rising too high.

Short-acting insulins must be injected regularly. They are now commonly administered using an insulin pump or as needed just prior to a meal. Insulin pumps deliver small, controlled doses on a regular basis to maintain glucose levels more constant and minimize glucose spikes that may cause protein glycosylation and tissue damage. Long-acting insulins (*insulin glargine*) are taken daily to maintain a baseline level of insulin in the blood. Prediabetes is treated with diet and lifestyle changes. In both type 1 and type 2 diabetes, regular monitoring of blood glucose levels is critical to ensure sufficient but not too much drug is being taken.

The first-line drug for type 2 diabetes is *metformin*. Metformin decreases hepatic glucose production by inhibiting glucagon signaling and increases insulin sensitivity in tissues by activating AMPK, 5'-AMP kinase, an important signaling molecule. The enzyme dipeptidyl peptidase 4 (DPP-4) breaks down incretins. DPP-4 inhibitors (*sitagliptin*) maintain high incretin levels that increase insulin secretion and suppress glucagon release. They are generally added to a metformin regimen and less often used alone.

Epinephrine and Hyperglycemia

Explain how epinephrine causes hyperglycemia in response to stress

Epinephrine mobilizes glucose and fatty acids for energy needs when the body is subjected to stress. Nearly all of the circulating epinephrine is provided by the adrenal medulla in response to sympathetic activation. Epinephrine inhibits insulin secretion and increases glucagon secretion. It supplements the effects of glucagon by stimulating hepatic glycogenolysis and adipose tissue lipolysis by activation of hormone-sensitive lipase. In muscle tissues, epinephrine increases anaerobic glycolysis and inhibits insulin-mediated glycogenesis. The end result is hyperglycemia.

> Epinephrine released in response to stress increases hepatic glycogenolysis and adipose tissue lipolysis to increase blood glucose and provide fatty acids as alternative substrates for energy

Glucocorticoids

Explain how glucocorticoids contribute to increased glucose

Glucocorticoids (cortisol) are released in response to long-term stress. They counteract the effects of insulin to maintain blood glucose levels for the brain. Glucocorticoids, like epinephrine and glucagon, increase hepatic gluconeogenesis and stimulate adipose tissue lipolysis. In peripheral tissues, they inhibit insulin-dependent glucose uptake and increase protein catabolism. The result of impaired glucose uptake and mobilization of glucose by the liver is to elevate blood glucose levels.

> Glucocorticoids (cortisol) increase glycogenolysis and lipolysis and make fat and muscle cells insulin resistant by blocking GLUT4 insertion which preserves blood glucose for brain during long-term stress

Ketone Bodies

Describe the production and utilization of ketone bodies

Ketone bodies (acetoacetate and beta-hydroxybutyrate) are short-chain derivatives of fatty acids and *ketogenic amino acids*. *Acetone* is generated from acetoacetate and may appear in the breath, giving it a fruity odor. Two amino acids are always ketogenic: leucine and lysine. Five others (phenylalanine, isoleucine, threonine, tryptophan, and tyrosine) can be either ketogenic (producing ketone bodies) or glucogenic (producing glucose). Ketone bodies are directly produced from acetyl-CoA, a two-carbon metabolic intermediate that ordinarily enters the citric acid cycle in mitochondria to produce energy by oxidative phosphorylation of ADP to ATP.

A small amount of ketone bodies is always produced. Normal blood levels are about one percent of glucose levels. Acetyl-CoA enters the citric acid cycle by combining with *oxaloacetate* to form citrate. When glucose levels are low in the liver, oxaloacetate is diverted into the gluconeogenic pathway and acetyl-CoA is used to produce ketone bodies. Such low glucose levels occur in fasting, starvation, low carbohydrate (ketogenic) diets, and prolonged strenuous exercise, which deplete hepatic glycogen. It also occurs in uncontrolled type 1 diabetes, due to lack of insulin and continuing stimulation of gluconeogenesis by glucagon. Similarly, when fatty acids are the preferred source of energy, the amount of acetyl-CoA produced exceeds the available oxaloacetate and acetyl-CoA is used to produce ketone bodies.

The brain preferentially uses glucose for energy. It does not require insulin for glucose uptake and is unable to use fatty acids (which do not readily pass the blood-brain barrier, anyway). The brain also uses ketone bodies for energy and to produce fatty acids to replenish cell membrane phospholipids. When blood glucose levels fall below normal levels, ketone bodies can be used to provide most of the energy needs of the brain. Thus, the production of ketone bodies is a protective response that maintains brain metabolism.

In uncontrolled diabetes, blood ketones and glucose can rise to dangerous levels. Both are spilled into the urine and can be detected there. Normal urinary ketones are < 5 mg/dL; values > 7 mg/dL (0.7 mM) are considered *ketonuria*. Ketonuria in non-diabetics is a sign of acute illness or severe stress due to elevated fat metabolism.

Ketone bodies are acidic. High ketone levels decrease blood pH (*ketoacidosis*). Sugars in the urine decrease the osmotic difference between the filtrate in nephron tubules and the surrounding interstitial fluids, which decreases water reabsorption by the kidneys, leading to dehydration. Thus, uncontrolled diabetes causes hyperglycemia, acidosis, and dehydration.

As glycogen stores are depleted in starvation, amino acids and fatty acids produce energy and ketone bodies that supply brain in place of glucose
Ketone bodies are acidic and may cause ketoacidosis with low carbohydrate diets

Metabolic Syndrome

Explain the relationship between metabolic syndrome and diabetes

A *syndrome* is a set of *signs* (that anyone can observe) and *symptoms* (reported by the patient) which occur together. *Metabolic syndrome* is a cluster of three of five medical conditions (signs): central obesity, hypertension, hyperglycemia, hypertriglyceridemia, low HDL levels. *Central obesity* is accumulation of fat in the abdominal region (visceral fat) as opposed to accumulation of subcutaneous fat.

The criteria for diagnosing the syndrome vary somewhat. Typically, 1) central obesity is defined as waist to hip ratio > 0.85 (female) or > 0.9 (male) or BMI (body mass index) > 30 kg/m^2; 2) systolic BP > 130 or diastolic > 85 mm Hg; 3) fasting blood glucose > 100 mg/dL; 4) triglycerides > 150 mg/dL; 5) HDL cholesterol < 50 mg/dL (females) or < 40 mg/dL (males).

Metabolic syndrome poses an increased risk for developing cardiovascular disease and type 2 diabetes. Incidence is increased with sedentary lifestyle, overweight, stress, and diets high in simple sugars. Central adiposity is commonly observed in most patients with metabolic syndrome. The first line of defense is a healthy diet (low in simple carbohydrates) and increased physical activity. However, compliance with such a regimen is problematic. Commonly, diuretics and ACE inhibitors are used to lower blood pressure, statins to improve blood lipid profiles, and metformin and sitagliptin for hyperglycemia.

Lactate and the Cori Cycle

Describe the mechanism by which lactate is used to generate glucose

Sources of metabolic energy for ATP synthesis in muscle depend on the level of activity. Resting muscle oxidizes fatty acids, produces creatine phosphate and stores glycogen. *Creatine phosphate* is a short-term energy store in muscle cells that directly generates ATP. Short bursts of high-intensity exercise rapidly use up the creatine phosphate stores. Once these are exhausted glycogen is hydrolyzed to glucose which enters glycolysis to produce ATP with lactate as a by-product (anaerobic metabolism).

In the *Cori cycle*, lactate goes to the liver and is converted back to glucose with ATP from fatty acid catabolism (gluconeogenesis). Since fatty acid oxidation requires oxygen, the lactate which is accumulated incurs an oxygen debt, which is made up after high intensity exercise is discontinued. *Lactate* may also be used as a source of energy in muscles and the brain. Because lactate moves from one tissue to another, this is called the lactate shuttle. Lactate produced by glycolytic muscle fibers engaged in high intensity exercise is shuttle to oxidative muscle fibers that are less active. High circulating levels of lactate are also used by the heart to produce energy for contraction.

Anaerobic metabolism of glucose produces lactate in skeletal muscles
Lactate carried to liver is used in gluconeogenesis (Cori cycle)
ATP for gluconeogenesis comes from fatty acid oxidation

Review Questions

1. What are nutrient pools? In what forms are nutrients carried in the blood?
2. What releases fatty acids from triglycerides at tissues?
3. How are fatty acids stored and released from adipose tissues?
4. Which tissues store glucose as glycogen?
5. Why is liver the only source of blood glucose from glycogen?
6. What is the response of insulin to elevated blood glucose?
7. What keeps blood glucose from falling too low?
8. What is the effect of insulin on adipose tissue and skeletal muscle fibers?
9. What is gluconeogenesis?
10. What are the effects of glucagon on hepatocytes and adipose tissues?
11. What are the effects of epinephrine on metabolism?
12. When are ketone bodies produced and how are they used?
13. How is lactate recycled in the liver?

Neurophysiology

The art of life is the art of avoiding pain; and he is the best pilot, who steers clearest of the rocks and shoals with which it is beset.

– Thomas Jefferson to Maria Cosway

Sensory Physiology

Nervous control can be divided into three main functional classes. The autonomic nervous system (along with endocrine hormones) regulates visceral functions. This has been the central theme of all that has gone before in these lectures. The *somatic nervous system* has two parts. Its sensory component monitors the external environment. Its motor component controls skeletal muscles.

In this lecture, the focus will be on *sensations*: how we sense the external environment, how these signals cause *spinal reflexes*, and the pathways from peripheral receptors to the brain. The primary concept to keep in mind is *labeled line coding*. The location and type of stimulus is maintained throughout the nervous system. Certain types of receptors respond to particular stimuli which are are sent to a part of the brain that interprets them as pain, or cold, or sound, or sight (*modality*). The pathways followed keep track of the location. Thus, the line is "labeled" with position and modality information.

This lecture describes the sensory part of the somatic nervous system. Many of the same principles apply to the special senses, but we do not discuss them here. Rather, the emphasis is on basic physiology of sensory systems and their integration into spinal reflexes. In the next lecture, we begin a two-part investigation of the the brain. After a structural introduction, we look at what it means to be awake and how the body deals with pain. The second part of our brain study zeroes in on how we make decisions about what to do.

Organization of the Nervous System

Distinguish between the two divisions of the nervous system and their subdivisions

The *nervous system* responds rapidly to stimuli, integrates signals, and activates muscles and organs. It consists of the brain, spinal cord, nerves, and sensory organs. The brain and spinal cord are the *central nervous system*. Nerves and sensory organs are the *peripheral nervous system*.

> Nervous system responds rapidly, integrates signals, activates muscles
> CNS (central nervous system): brain and spinal cord
> Peripheral nervous system: nerves and sensory organs

The peripheral nervous system has three divisions (somatic, autonomic, and enteric). The *somatic nervous system* carries neurons to and from the skin, tendons, and skeletal muscles (soma). The *autonomic nervous system* (ANS) controls the viscera (organs). The *enteric nervous system* is within and locally controls the GI tract. It is often included as part of the ANS.

The ANS has two divisions (sympathetic and parasympathetic). The *sympathetic division* of the ANS arises from thoracic and lumbar spinal nerves (thoracolumbar). The *parasympathetic division* of the ANS lies arises from cranial and sacral nerves (craniosacral).

Sensory Receptor Classification

Classify sensory receptors according to location and stimulus modality

General senses originate in many places, all over the body. The general senses include pain, heat, touch, and pressure. The *special senses* (olfaction, vision, gustation, audition, equilibrium) arise from specialized organs.

Exteroceptors sense the external environment. *Interoceptors* sense the internal environment. Both types respond to various stimuli (*modalities*) to a greater or lesser extent. *Chemoreceptors* respond primarily to chemicals. *Thermoreceptors* respond to changes in temperature (cold, hot). *Nociceptors* respond to pain. *Photoreceptors* respond to light and are found only in the retina. *Mechanoreceptors* respond to deformation (touch, pressure). They include *baroreceptors* that respond to pressure in internal organs and blood vessels and *proprioceptors* that monitor the position of muscles (muscle spindles), tendons (Golgi tendon organs) and joints.

> General senses originate from receptors all over the body
> Receptors respond to chemicals, temperature, pain, light, and deformation

Sensory Neurons

Describe how sensory neurons respond to stimuli

Sensory neurons for the general senses are *pseudounipolar*. Their cell bodies lie off to the side of their axon in the *dorsal root ganglia* just outside the spinal cord. They do not have dendrites. Action potentials begin in a sensory ending that is continuous with the axon in the periphery. Often this sensory neuron is encapsulated in a connective tissue capsule or is associated with a specialized connective tissue structure. Most have myelinated axons that transmit signals rapidly from the periphery to the central nervous system. Others are poorly myelinated and transmit signals more slowly.

> Sensory neurons for general senses are pseudounipolar neurons with cell bodies to the side of the axon and located in dorsal root ganglia
> Receptors may be free or encapsulated nerve endings or separate cells

Sensory nerve endings respond to a stimulus by initiating action potentials in their axons. Pressure, or binding of a chemical, or changes in temperature open ion channels that allow Na ions to enter the sensory ending. This causes a graded depolarization (*receptor potential*) that spreads from the sensory ending to the *initial segment* of the axon where voltage-gated channels are located. As in all axons, if the graded potential reaches threshold, *voltage-gated Na channels* in the axon open and initiate an action potential that travels the entire length of the axon and into the central nervous system. This sensation must then be interpreted by the CNS to give the *perception* of a general sense, such as touch, heat, or pain.

The intensity of a stimulus is encoded in the *frequency* of action potentials. Stronger stimuli cause action potentials to fire with greater frequency. This leads to a release of a larger amount of neurotransmitter at the axon terminal in the CNS. The extent of the response is time-dependent. *Receptor adaptation* refers to a decrease in sensitivity to a continued stimulus. *Phasic receptors* are rapidly adapting; they signal a change in stimulus (at on and off). *Tonic receptors* are slowly adapting and respond for the duration of a stimulus.

Receptive Fields

Define receptive field, acuity, two-point discrimination, and lateral inhibition

A *receptive field* is the site over which a receptor responds to a signal. *Acuity* is the ability to precisely localize closely spaced stimuli and depends on receptive field size. Receptive fields in areas of fine touch (fingertips) are very small; those for crude touch are larger. With overlapping receptive fields, nearby signals merge and are sensed as a single stimulus. With isolated receptive fields, the stimulus is sensed as two separate signals (*two-point discrimination*).

> Receptive field is area over which a receptor responds to a stimulus
> Acuity: ability to discriminate closely spaced stimuli
> Two-point discrimination: distance between points perceived as separate

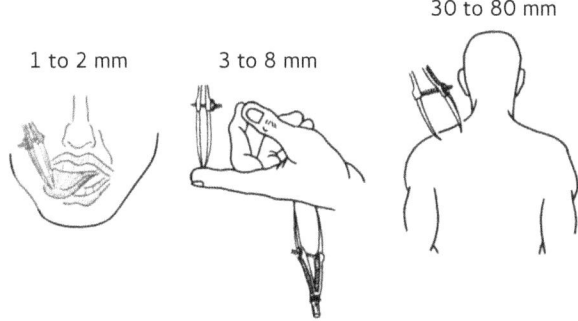

More action potentials are produced in the receptor closest to the stimulus than those around it. Signals at the edge of the receptive field may be inhibited to sharpen response (*lateral inhibition*) by reducing signals from nearby neurons. This inhibition is the result of inhibitory neurons that reduce neurotransmitter release from the surrounding neurons in the CNS.

> In lateral inhibition, signals from the receptor closest to the stimulus
> inhibit action potentials produced by neighboring receptors
> to sharpen the output signal by reducing surrounding noise

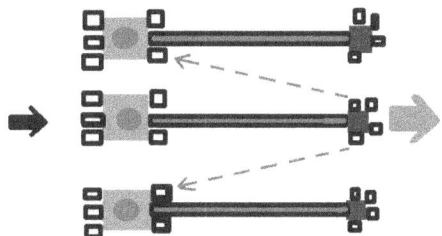

Skin Receptors

Describe the general sensory receptors of the skin and hypodermis

The *skin* is a protective barrier that separates the warm and wet inside of the body (the tissues) from the cold and dry outside environment. Outside is all scary and dangerous; inside is cozy and safe. Receptors in the skin and special senses constantly monitor the external environment, looking for dangers to avoid.

Skin areas innervated by sensory fibers of a single spinal nerve are *dermatomes*. Sensory neurons of the skin respond to pain, pressure, touch, and chemicals. *Free nerve endings* are nociceptors that sense pain. The *root hair plexus* surrounds the base of hairs and senses brushing motions on the skin. Thermoreceptors are also free nerve endings. Warm receptors respond with higher action potential frequency upon warming and lower frequency signals when cooled. Cold receptors respond in the opposite fashion, but also produce an a paradoxical rapid, brief discharge when temperature increases above 45C (110°F).

> Skin receptors respond to pain (free nerve endings), brushing (root hair plexus), touch (Merkel receptors), vibrations (Meissner's corpuscles), pressure (Pacinian corpuscles), and stretch (Ruffini corpuscles)

Merkel receptors are associated with Merkel cells in the stratum basale and respond to fine touch and pressure. They have small receptive fields. When the Merkel cell is disturbed by movement of the epidermis it produces chemicals that act on the sensory ending of a closely associated sensory neuron. This generates a sustained response (string of action potentials) so long as the stimulus is present. Thus, they are classified as slowly adapting mechanoreceptors. Merkel receptors are abundant in the fingertips and can detect the edges of objects.

> Merkel receptors respond to chemicals released by closely associated Merkel cells that lie in basal layer of epidermis and sense light pressure They are slowly adapting, abundant in fingertips, and detect object edges

A *Meissner corpuscle* is an egg-shaped mass of sensory nerve endings enclosed by a connective tissue capsule. Meissner corpuscles are in the dermal papillae of hairless skin that extend upward into the epidermis. The connective tissue capsule of a Meissner corpuscle shields its nerve endings from direct pressure. These receptors have small receptive fields and are most abundant in the finger tips. Meissner corpuscles are rapidly adapting, responding to the beginning and end of a stimulus. They primarily sense light touch and vibrations, so are often called *tactile corpuscles*.

Meissner corpuscles are a mass of sensory endings enclosed in a connective
tissue capsule in dermal papilla of hairless skin that detect vibrations
They are rapidly adapting and respond to onset and end of light pressure

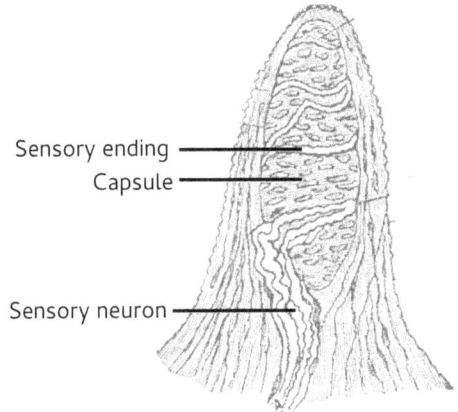

Sensory ending

Capsule

Sensory neuron

A *Pacinian corpuscle* is a multilayered connective tissue capsule surrounding a single sensory ending in the dermis. Its 20-60 layers give
rise to the alternative name of *lamellar corpuscle*. Deformation of the
lamellae pushes on the sensory ending to cause a receptor potential that
is transferred to the axon of the sensory neuron to generate an action
potential. Pacinian corpuscles are sensitive to changes in pressure and
vibrations that allow them to assess the texture of an object's surface
(rough or smooth). They have large receptive fields and produce transient (phasic) responses. This rapid adaptation prevents Pacinian
corpuscles from constantly responding to the friction of clothing.

Pacinian corpuscles in the dermis are sensory endings surrounded by
multiple layers of connective tissue that must be compressed to respond
They are rapidly adapting and detect textures of objects

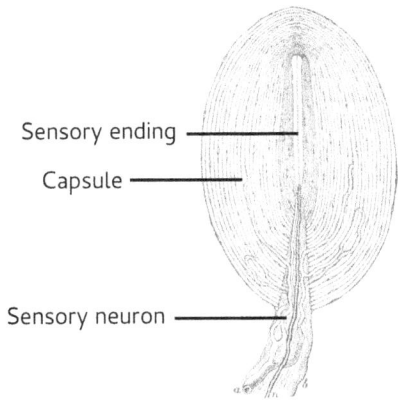

Sensory ending

Capsule

Sensory neuron

A *Ruffini corpuscle* is an elongated, encapsulated receptor deep in the dermis. The connective tissue surrounding the receptor is closely integrated into the surrounding dermal tissue. Ruffini corpuscles are sensitive to stretching of the dermis during movement. Similar receptors are observed in ligaments and tendons. Ruffini corpuscles have large receptive fields and are slowly adapting, producing sustained responses to constant stimulation. Their density is highest around the fingernails. Thus, they are important in control of finger movements and in monitoring the slippage of objects held in the fingers. That is, they ensure a good grip on grasped objects.

> Ruffini corpuscles have diffuse sensory endings that
> course through surrounding connective tissue of dermis
> They are slowly adapting and monitor slippage of objects held by fingers

Spinal Cord

Describe the anatomy of the spinal cord and spinal nerves

The spinal cord lies within the vertebral column. Like the brain, it is protected by bone and a set of connective tissue layers, the *meninges. Meninges* are protective membrane (connective tissue) coverings lying beneath the bone. *Dura mater* (hard mother) is the outermost, fibrous covering of the meninges. *Arachnoid mater* (spider mother) is a delicate fibrous membrane attached to the dura mater. It sends *arachnoid trabeculae* down to the underlying pia mater forming the *subarachnoid space* through which *cerebrospinal fluid* (CSF) circulates. CSF bathes brain and spinal cord to provide cushioning and nutrients, while removing wastes. This delicate network looks like a spider web, thus the name. The arachnoid mater does not enter the convolutions of the brain. *Pia mater* (tender mother) is the innermost covering of brain and spinal cord that dips into ridges and clefts, firmly adhering to all the contours of the brain and spinal cord. Its capillaries provide nourishment to the underlying nervous tissue.

The *spinal cord* has gray matter in its central horns, which have a butterfly shape. Signals travel along white matter tracts (columns) and communicate by synapses in gray matter. *Gray matter* includes neuronal cell bodies, dendrites, unmyelinated neurons, and axon terminals. *White matter* (tracts of myelinated axons) in the spinal cord surrounds the gray matter. White matter carries signals up (ascending) and down (descending) the spinal cord.

Sensory neurons enter the spinal cord on its dorsal side and synapse in the gray matter. Since these pseudounipolar neurons carry signals to the central nervous system (CNS), they are *afferent neurons*. Cell bodies of sensory neurons lie outside the spinal cord in dorsal root ganglia. *Motor neurons* exit on the ventral side of the spinal cord and have their cell bodies in spinal cord gray matter. Since these multipolar neurons carry signals away from the CNS, they are *efferent neurons*.

> Spinal cord is protected by vertebra and meninges
> Spinal nerves contain both sensory and motor neurons
> Sensory neurons enter dorsal roots; motor neurons exit ventral roots

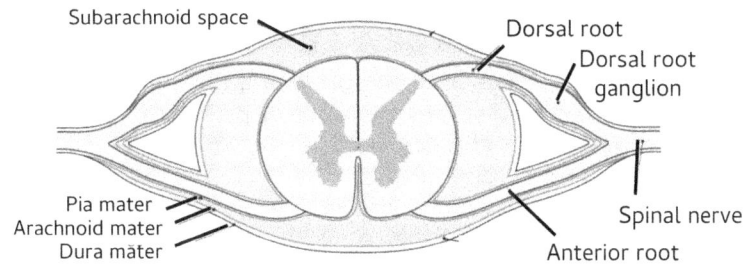

The spinal cord ends in the *conus medullaris* at the level of lumbar vertebra L1-L2 because it does not continue to elongate after infancy. Spinal nerves continue downward in the *cauda equina* (horse's tail) to exit through lower lumbar and sacral vertebrae. Outside the cauda equina is the subarachnoid space, filled with CSF that can readily be sampled by lumbar puncture at L3-L4 without risk of injury to the spinal cord. Outside the dura mater of the spinal cord, but still within the vertebra is the *epidural space*, which is filled with fat and serves as a site for anesthesia.

> Epidural space is outside dura mater but within vertebrae
> Spinal cord ends at L1-L2; nerves continue as cauda equina
> Subarachnoid space around cauda equina can be sampled for CSF

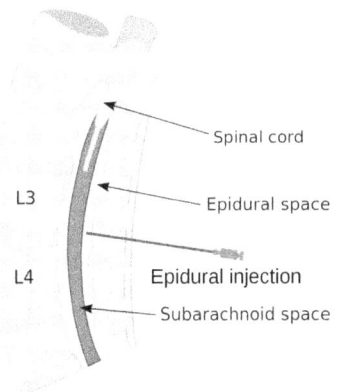

Labeled Line Coding

Explain how position and mode are maintained throughout somatosensory pathways

All sensory stimuli project to specific areas of the brain (*sensory transduction*). Depending on the type and location of the stimulus, the brain interprets the signal (*perception*). *Labeled line coding* maintains information on the type and location of sensations throughout nervous system. The *posterior column pathway* carries signals for discriminative touch (distinguishing two closely spaced points), pressure, vibration, and proprioception to the cerebral cortex of the brain. The *spinothalamic tract* carries sensations for crude touch, pressure, and pain to the cortex.

> Cell bodies of sensory (afferent) neurons are in dorsal root ganglia
> Sensory neurons enter dorsal roots to synapse in posterior gray matter
> Motor (efferent) neurons exit spinal cord through ventral roots

Sensory pathways (ascending tracts) pass through three neurons (first-, second-, and third-order). *First-order neurons* are sensory neurons in the periphery that enter the spinal cord. *Second-order neurons* travel up the spinal cord, *decussate* (cross over) to the opposite side, then project to the thalamus. Sensory neurons of the posterior column pathway decussate in the brainstem; those of the spinothalamic tract cross through the gray commissure at their site of entry into the spinal cord. This decussation maps the right side of the body to the left side of the somatosensory cortex. *Third-order neurons* travel from the thalamus to specific regions of the sensory cortex (except for olfaction). There, they are mapped as a little person (*sensory homunculus*) in proportion to sensory sensitivity of the site of origin.

> Sensory signals ascend to brain through specific white matter tracts
> Most sensations (touch, texture) pass up posterior column
> Position and modality is retained throughout (labeled line coding)

Proprioception

Describe the receptors for body position

Cutaneous mechanoreceptors primarily sense the external environment. They play some role in our sense of body position, but the primary receptor types involved in movement are the *proprioceptors* ("self" receptors) of joints, tendons, and skeletal muscles. Their output is closely integrated with that from the vestibular receptors for equilibrium in the inner ear. Fibrous capsules in the joints are similar to those found in the skin. Their role in proprioception is minimal.

At the site where muscle fibers merge into tendons are Golgi receptor organs that sense overall muscle stretch. However, most of our sense of position (kinesthetic sense) is obtained from the complex muscle spindle receptor within skeletal muscles. Muscle spindles are most abundant in the muscles that control precise movements, such as those of the hand and around the eye.

Golgi tendon organs lie at the junction of skeletal muscles with tendons
Nerve fibers enter a capsule that contain sensory endings
sensitive to stretch that wrap around collagen fibers of the tendon

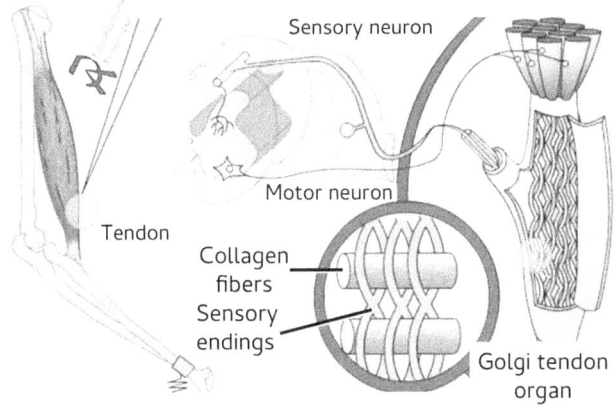

Sensory endings of *Golgi tendon organs* spiral around strands of collagen fibers that connect the end of a muscle to its associated tendon. They are surrounded by a fibrous tissue capsule. The axon of the sensory neuron is a large diameter, myelinated Ib fiber. Typically, several nerve fibers penetrate the capsule, where they lose their medullary sheaths. Muscle contraction pulls on the stiff tendon, slightly deforming the collagen fibers and the associated sensory endings. This opens stretch-activated cation channels to produce a receptor potential (depolarization) that spreads to the axon to generate an action potential that travels to the spinal cord.

Within the tissue of the muscle itself are encapsulated receptors (*muscle spindles*) that respond to muscle position and stretch. Inside the capsule are intrafusal muscle fibers that run parallel to the (extrafusal) skeletal muscle fibers that cause muscle contraction. Sensory endings wrap around the intrafusal fibers within the capsule. Together these sensory endings respond to changes in muscle length and velocity. The primary receptor projects to the spinal cord as large, myelinated type Ia (Aα) axons. A secondary receptor carries signals along type II (Aβ) sensory fibers. The output signals from muscle spindles and Golgi tendon organs pass to the cerebellum through the spinocerebellar tract that lies along the lateral edge of the spinal cord.

The ends of intrafusal fibers of muscle spindles are contractile. They contract in response to signals from gamma motor neurons which are activated at the same time as alpha motor neurons that cause contraction of the surrounding muscle. If a muscle is stretched, muscle spindles stretch along with it. Stretch activates sensory fibers in the muscle fiber and sends signals to the CNS that the muscle is not in its intended position.

> Muscle spindles are sensory endings that surround intrafusal muscle fibers
> Gamma motoneruons match intrafusal fiber tension to surrounding muscle
> Stretching muscle stretches intrafusal fibers and sends signals to spinal cord

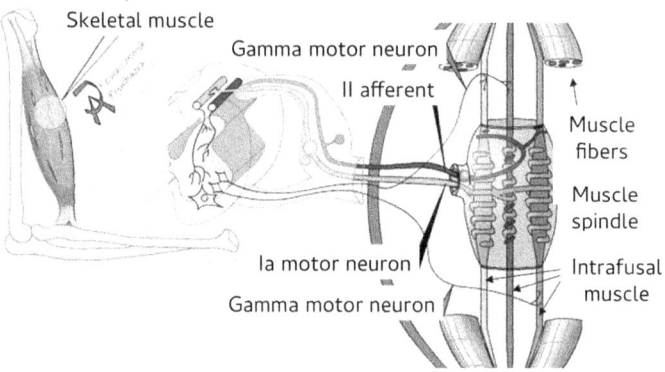

Spinal Reflexes

Describe the overall wiring of a spinal reflex

Reflexes are the primary functional response of the nervous system. An innate reflex is unlearned, automatic, and involuntary. A learned reflex is acquired through practice and repetition. A *reflex arc* is the wiring of a reflex. It consists of a *receptor* that sends signals through an afferent neuron to an *integrating center* in the CNS that responds by sending signals through efferent (motor) neurons to *effectors*, such as muscles or glands. Reflexes may be *somatic reflexes* (with output to skeletal muscles) or *visceral reflexes* (output to body organs).

Spinal reflexes are automatic, involuntary responses that are integrated in the spinal cord. Cranial nerve reflexes pass through cranial nerves and are integrated in the brainstem. Sensory (afferent) neurons are pseudounipolar and carry signals to the CNS. Specialized nerve endings in sensory neurons (receptors) respond optimally to a particular kind of stimulus. Motor (efferent) neurons carry signals to effectors (muscles, glands, and organs). *Interneurons* connect together neurons in the CNS and form the link between motor and sensory neurons. Motor neurons and interneurons are multipolar.

> Spinal reflex arc has peripheral receptor that sends signals to spinal cord
> In spinal cord, signal is passed to motor neurons of appropriate muscles
> Afferent neurons go to CNS; efferent neurons go from CNS to effectors

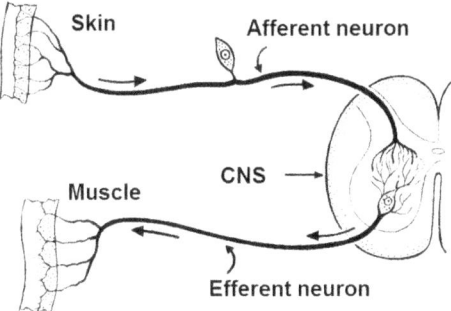

Monosynaptic Stretch Reflex

List the components of a monosynaptic reflex and describe the stretch reflex

In a *monosynaptic reflex*, sensory neurons send signals directly to motor neurons without any intervening interneuron. The (monosynaptic) *stretch reflex* returns a muscle to its proper length when stretched. Axons from muscle spindles synapse with alpha motor neurons in the spinal cord that innervate the contractile muscle fibers surrounding the sensory receptor of the *muscle spindle*. Activation of these neurons causes a stretched muscle to contract. Muscle spindle sensory neurons also synapse with inhibitory neurons that cause relaxation of the antagonist muscle to allow movement to take place. Signals are also sent along the spinocerebellar tract to the cerebellum that monitors body position.

When the *patellar ligament* is stretched with a reflex hammer, the tendon pulls on the muscle fibers, causing them to stretch. This activates muscle spindles, which send signals to the spinal cord. Alpha motor neurons to the quadriceps are excited, causing the muscle to contract and restore the intrafusal fibers of the muscle spindle to their proper length. At the same time, signals to the antagonist hamstring flexors are inhibited. This inhibition of the antagonist muscle is *reciprocal inhibition*. The result is a reflex kick of the foot. Note that the patellar ligament is often incorrectly called the patellar tendon. It connects the tibia to the patella, so it is a ligament (connects bone to bone) and not a tendon (that connects bone to muscle).

> A monosynaptic reflex directly connects muscle spindle Ia sensory neuron output to an alpha motor neuron that innervates the surrounding muscle
> Stretching the patellar ligament causes a reflex kick as quadriceps contracts

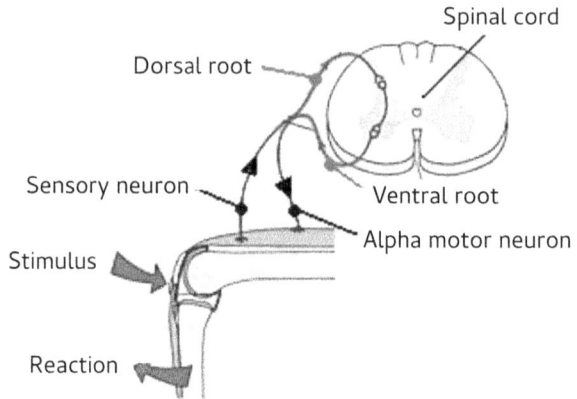

Golgi Tendon Reflex

Describe the Golgi tendon reflex

Activation of the *Golgi tendon organ* by stretch causes relaxation of the associated muscle. Since the Golgi tendon organs are organized in series with the extrafusal muscle fibers that cause contraction (that is, in the tendons), they stretch when the muscle contracts. This is distinct from the muscle spindles, which are parallel to the extrafusal muscle fibers of the contracting muscle (they are buried within the muscle) and report directly on their length. While activation of muscle spindles by stretch causes muscle contraction (*myotatic reflex*), activation of Golgi tendon organs by stretch causes muscle relaxation (*inverse myotatic reflex*).

Sensory neurons from the Golgi tendon organ synapse on inhibitory neurons that decrease the rate of action potential firing by the alpha motor neurons that innervate their associated muscle. This limits how strongly a muscle can contract and prevents damage to tendons and joints. Since two synapses are involved, this is a *disynaptic reflex*.

> The Golgi tendon reflex causes muscle relaxation in response to muscle contraction that stretches the associated tendon
> Ib afferent activates inhibitory interneuron that inhibits alpha motor neuron

Polysynaptic Reflexes

Distinguish between a polysynaptic and monosynaptic reflex and describe the withdrawal reflex

In a *polysynaptic reflex,* one or more interneurons coordinate a response at a distance from the stimulus. Sensory input excites an excitatory interneuron that synapses on a motor neuron to the agonist. The alpha motor neuron for the agonist does not necessarily arise from the same dermatome as the input stimulus. The neuronal connection may need to pass up or down the spinal cord to reach the alpha motor neuron of the correct prime mover.

At the same time, an inhibitory interneuron links the input from the sensory neuron to reduced firing of the alpha motor neuron supplying the ipsilateral antagonist muscle. Both of these connections are on the same side of the spinal cord. The result is contraction of the prime mover and relaxation of its antagonist by *reciprocal inhibition.* Reciprocal inhibition is also necessary in the patellar tendon reflex in order to relax the hamstrings as the quadriceps femoris contracts.

An example of a polysynaptic reflex is the *withdrawal (flexor) reflex* that rapidly removes a body part from a source of pain, such as by touching a hot stove. The biceps brachii and brachialis are excited while the triceps brachii relaxes to cause elbow flexion, removing the hand from danger. Clearly, the muscles involved are distant and distinct from the source of the painful stimulus. At the same time, motor units across the spinal cord are recruited to stabilize posture on the contralateral side of the body. Extensors are activated and flexors are inhibited. This part of the withdrawal reflex is the *crossed extensor reflex.*

> Withdrawal reflex is polysynaptic: interneurons direct response to distant muscles; prime mover contracts and antagonist reflexes
> Crossed extensor reflex stabilizes contralateral side in withdrawal reflex

Review Questions

1. Distinguish between general and special senses, interoceptors and exteroceptors.

2. What are the five classes of sensory receptors? What are two types of mechanoreceptors?

3. List and describe the functions of the six receptors in the integument.

4. What is a receptive field? How is acuity improved by lateral inhibition?

5. How do tonic and phasic receptors differ? How is stimulus intensity encoded in neurotransmitter release?

6. Describe the structure of the spinal cord, gray and white matter, dorsal and ventral roots.

7. What are the three layers of the meninges in brain and spinal cord?

8. Describe the production, flow, and function of cerebrospinal fluid (CSF).

9. Distinguish between epidural space and subarachnoid space.

10. What is labeled line coding? What are the three neurons in somatosensory pathways?

11. List the functions of the posterior and spinocerebellar columns and spinothalamic tract.

12. What is proprioception? What receptors monitor body position?

13. How does a monosynaptic reflex differ from a polysynaptic reflex?

14. What is the Golgi tendon reflex?

15. Why are reciprocal inhibition and the crossed extensor reflex required in a withdrawal reflex?

Consciousness and Pain

In this first of two lectures on the brain, we look into its basic anatomical organization into *six divisions* and how it is supplied with nutrients by the *cerebrospinal fluid* while being protected from toxic substances by the *blood-brain barrier*. We study the basic substructure of the brain: its medullated *tracts* that carry signals from one region to another and its *nuclei*, where synaptic connections are made.

Electrical signals in the brain can be recorded by an *electroencephalogram* (EEG). These recordings change significantly over a 24-hour period (the *circadian rhythm*) as we cycle between states of wakefulness and sleep. Wakefulness is governed by the *reticular activating system* in the brainstem that sends signals throughout the brain that keep us from falling asleep while we are engaged in activity.

The function of the brain is determined by how its parts (the nuclei) are connected by synapses and the type of signals that pass from one neuron to another across the synaptic cleft. These signals may be excitatory or inhibitory depending on the neurotransmitter and the receptors to which it binds. Any one postsynaptic neuron receives multiple inputs that may or may not lead to the firing of an action potential. *Excitatory* signals make it more likely for an action potential to fire. *Inhibitory* ones make it less likely to fire.

To better understand how these signals are integrated, we will briefly study the mechanism of action of anxiolytics, analgesics, and anesthetics. Anxiolytics are also sedatives that can be used to induce sleep or cause relaxation before surgery. Most increase signaling by the inhibitory neurotransmitter *GABA*.

We spend some time examining *pain* mechanisms, the natural pain suppressing response of the body, and the role of opioids in reducing pain. We see how specific pathways and neurotransmitters are involved with pain and how these may be modified with selective drugs. Analgesics may act in the periphery as well as in the brain. In the CNS, analgesics and anesthetics increase inhibitory *GABA* signaling while reducing excitatory *glutamate* signals.

Brain

List the divisions of the brain and their functions

The *brain* receives sensory input, integrates information, makes decisions, and initiates motor activities. It is not an independent organ but is intimately connected with the spinal cord and the periphery. The brain has six divisions: medulla oblongata, pons, midbrain, cerebellum, diencephalon, and cerebrum. The three divisions closest to the spinal cord (medulla oblongata, pons, midbrain) are the *brainstem*. The *diencephalon* is in the middle of the brain and consists of thalamus and hypothalamus. The brainstem and *hypothalamus* are responsible for most autonomic functions. For example, the *medulla oblongata* contains centers that control respiration, blood pressure, and heart rate.

Medulla oblongata, pons, midbrain form brainstem
Diencephalon: thalamus (relay centers) and hypothalamus (visceral control)
Cerebellum coordinates movement and cerebrum controls higher functions

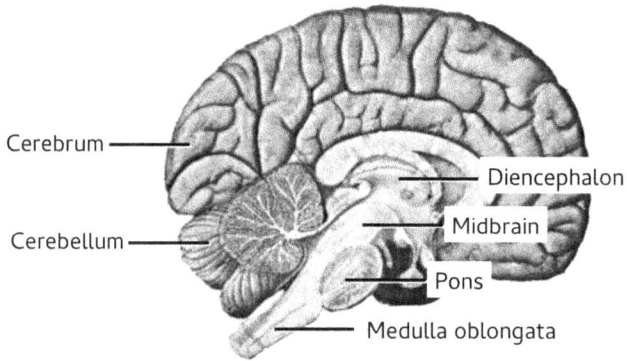

The *thalamus* is a relay center for all sensory input (except olfaction) and motor fibers. The *cerebellum* controls posture and monitors and adjusts movements to match intentions. The *pons* connects the upper and lower regions of the brain to the cerebellum. The *cerebrum* is the seat of intelligence. It processes sensations, plans, and initiates movements. It rests on *cerebral peduncles* of the *midbrain* that carry its ascending sensory and descending motor fiber tracts.

Blood Brain Barrier

Describe the functions of astrocytes and their role in forming the BBB

Astrocytes provide nutrients to neurons and remove excess ions and neurotransmitters after neural firing. Astrocyte processes contact blood capillaries, neurons, and pia mater. Astrocytes form a scaffolding for neural migration during development. They form a *tripartite synapse* with the axon terminal of a presynaptic neuron and the postsynaptic cell. There, they contribute to recycling of neurotransmitters and influence their concentrations in the synaptic cleft.

> Blood brain barrier consists of tight junctions between endothelial cells that prevent transfer of potential neurotoxins from blood to brain
> Those that enter cells are pumped back out by P-glycoprotein

When in contact with capillaries, astrocytes help maintain the *blood-brain barrier* that prevents dangerous substances from entering the brain. This barrier is formed by tight junctions between the cells of the capillary endothelium. Small, hydrophobic molecules, including oxygen and carbon dioxide, readily pass across the endothelium by passive diffusion through the cell membranes. Glucose and amino acids pass through by selective transporters. Lipophilic molecules that are potentially neurotoxins are actively transported out of the endothelial cells by a membrane-associated *P-glycoprotein*.

Cerebrospinal Fluid

Describe the production and circulation of cerebrospinal fluid

Cerebrospinal fluid (CSF) is a clear liquid, with a composition similar to that of plasma It is produced in the *choroid plexus* in the ventricles of the brain. Unlike plasma, it is nearly free of proteins. It cushions and protects the brain. From the ventricles, CSF passes into the subarachnoid space to circulate around the brain and spinal cord. It returns to the venous circulation through *arachnoid granulations* that extend into the venous sinuses of the brain. Valves in arachnoid granulations ensure one-way flow.

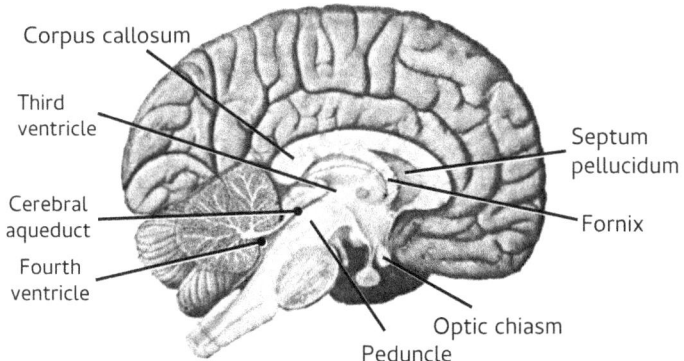

Cerebrospinal fluid (CSF) produced by choroid plexus of brain ventricles
circulates through subarachnoid space of brain and spinal cord
CSF returns to venous circulation through arachnoid granulations

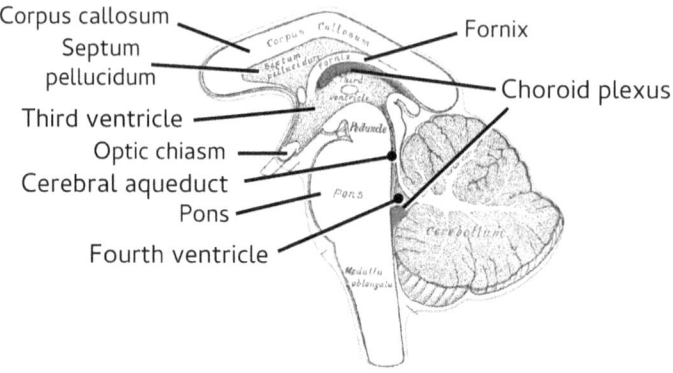

Nuclei and Tracts

Distinguish between nuclei and tracts and describe their roles

In the CNS, neurons are myelinated by *oligodendrocytes*. In the spinal cord, myelinated white matter tracts that carry signals up and down the cord surround gray matter containing cell bodies and synaptic connections. In the brain, white matter tracts are centrally located. Gray matter is on the surface (cortex) of cerebrum and cerebellum and in distinct clusters within the interior of the brain. A *nucleus* is a site of gray matter in the CNS; a *ganglion* is a cluster of gray matter in the peripheral nervous system. Traditionally the prominent nuclei in the central regions of cerebrum (caudate, putamen, and globus pallidus make up the *basal ganglia* but are more properly called *basal nuclei*. The thalamus is another prominent nucleus.

Projection fibers are white fiber (medullated) tracts that pass upward in brain
Commissural fibers (corpus callosum) connect left and right hemispheres
Nuclei are regions of gray matter containing cell bodies and synapses

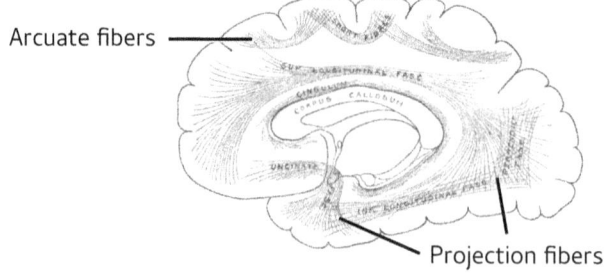

The cerebrum is divided into a left and right hemisphere by the longitudinal fissure. The right and left sides of the brain are connected by white matter tracts (the *corpus callosum*). In addition to this *commissural fiber* tract, myelinated white *projection fiber* tracts connect lower and higher centers and *arcuate fibers* link adjacent gyri. Major projection fibers (*internal capsule*) carry signals around the basal nuclei in the center of the brain and project outward to the cerebral cortex (*corona radiata*).

EEG and Imaging

Describe the measurement of an electroencephalogram

An *electroencephalogram* (EEG) records small electrical signals that arise from brain activity. It is used to diagnose epilepsy, monitor sleep patterns, and to monitor brain function in the ICU. Focal brain disorders, such as stroke, are now more commonly diagnosed with imaging techniques using x-rays (CT, *computed tomography*/PET, *positron emission tomography*) or NMR (MRI, *magnetic resonance imaging*).

In *MRI*, a strong magnetic field surrounds the region to be investigated. This excites hydrogen nuclei (mostly of water), which emit a radio frequency signal detected by the instrument. The magnetic field is switched on and off and the rate of relaxation of the back to ground state is recorded. A *CT scan* processes multiple x-rays to generate a set of cross-section images (tomography). Radiocontrast agents (usually iodine-based) are injected to improve image quality. *PET* is a type of tomography that detects gamma rays produced by an injected tracer molecule, typically radioactively fluorinated glucose (fludeoxyglucose). It maps metabolic activity so is a type of *functional imaging*. The two methods are sometimes combined in a PET-CT scan.

> MIR (magnetic resonance imaging) detects signals from hydrogen nuclei
> CT collects multiple x-rays to form a series of cross-sectional images
> PET uses stable isotopes to monitor metabolic activity

EEG waves are separated out according to their frequency range (how fast they oscillate), measured in Hz (cycles per second). *Alpha waves* (8-13 Hz) are the normal waveform with eyes closed, but not asleep. They disappear with attention or concentration on a task, such as doing mental arithmetic with the eyes open. *Beta waves* (faster than 13 Hz) increase with active thinking, anxiety, and focus. *Theta waves* (3.5 to 7.5 Hz) and *delta waves* (3 Hz or less) are slow waves, characteristic of sleep. Theta waves are observed when drowsy. Delta waves are associated with deep sleep.

In a typical EEG, multiple electrodes are placed on the surface of the scalp. Signals from distinct regions distinguish focal (localized) abnormalities associated with partial (focal) epilepsy from generalized epilepsy. The most common location of partial seizure abnormalities is the temporal lobe (just in front of the ear, under the temporal bone). Typically, an EEG is not recorded during an epileptic seizure but between seizures (*interictal epileptiform discharges*, IEDs). Rapid spikes indicate abnormalities, but are difficult to distinguish from artifacts. Intermittent photic stimulation (flashing lights) are often used to elicit IEDs.

> Electroencephalogram (EEG) records electrical activity of brain using
> surface electrodes; data analyzed into waves of different frequencies

Circadian Rhythms

Describe the regulation of circadian rhythms

Circadian rhythms are daily cycles during which the levels of hormones (and other biological activities) change throughout the day. Light-dark cycles are monitored by specialized cells in the retina that are photosensitive (respond to light) but are not involved in vision. These *photosensitive retinal ganglion* cells send signals through the retinohypothalamic tract to the *suprachiasmatic nucleus* (SCN) of the hypothalamus, directly above the optic chiasm.

> Photosensitive retinal ganglion cells synchronize clock in
> suprachiasmatic nucleus (SCN) of hypothalamus to light-dark cycle
> SCN sends signals through multineuronal pathway to pineal gland

Activities of circadian clock proteins in the SCN rise and fall in a *diurnal rhythm* of close to 24 hours. Input from photosensitive retinal ganglion cells shifts this rhythm to correspond with the actual light-dark cycle. The SCN sends inhibitory signals to the nearby paraventricular nucleus (PVN) which is next to the third ventricle. SCN firing rate is maximal mid-day, which keeps output from the PVN inhibited. The PVN sends neurons down the spinal cord then back through the superior cervical ganglia to the pineal gland, which is just posterior to the thalamus.

The *pineal gland* secretes *melatonin* when stimulated. At night, decreased SCN activity releases its inhibition of the pineal gland through the PVN. Thus, secretion of melatonin peaks at night and ebbs during the day. Melatonin organizes *sleep-wake cycles* and regulates core body temperature, which peaks in the evening and is lowest in early morning, before waking. Plasma cortisol levels also show a circadian rhythm: they peak mid-morning (increasing energy metabolism), then decrease at night.

Sleep

Describe the stages of sleep

Sleep is a state of altered consciousness with reduced ability to react to stimuli. It occurs in cycles of about 90 minutes each. *NREM* (non-rapid eye movement) sleep has multiple phases. All are dreamless and muscles are not paralyzed. In Stage 1 (light sleep), alpha waves disappear and theta waves appear. This is the entry level of the sleep cycle. Stage 2 NREM shows characteristic sleep spindles (a burst of 12-14 Hz waves). In both stages, the sleeper is easily aroused. The deepest phase (Stage 3, deep sleep) has at least 20% delta wave activity (*slow wave sleep*). Slow wave sleep improves declarative memory by facilitating memory consolidation. The brain is quite active. Bedwetting, night terrors, and sleepwalking occur during slow wave sleep.

REM (rapid eye movement) sleep is a time of dreaming and muscle paralysis that follows slow wave sleep. The EEG has rapid, low-voltage, desynchronized waves. It is paradoxical in that the brain waves are similar to those of the awake state, even though the person is asleep and difficult to arouse. Eye movements are rapid compared with other phases of sleep, but not as rapid as in an awake individual. Temperature control is compromised, so hot or cold environments reduce the proportion of REM sleep. It is best to awake from light sleep (NREM Stage 1), after having passed through a period of REM.

> Sleep occurs in cycles,each consisting of several stages
> NREM (non-rapid eye movement) sleep: increased slow delta waves
> REM (rapid eye movement): difficult to arouse, wakeful EEG pattern

Slow wave sleep

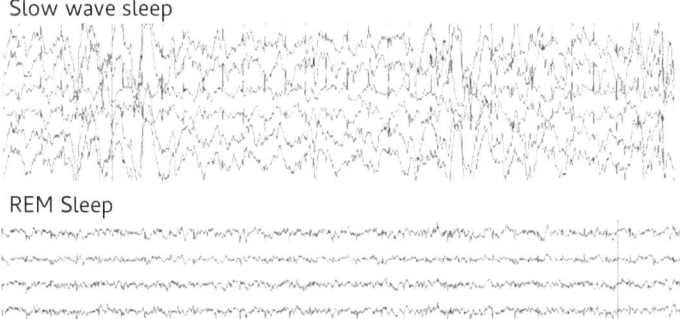

REM Sleep

Transition from one stage of sleep to another can be monitored by simultaneous recordings of brain waves (EEG), skeletal muscle activity

(EMG, electromyogram), and eye movements (EOG, electooculogram). Such a recording is a *hypnogram*. Additional sensors for EKG, nasal and oral airflow, pulse oximetry (which measures arterial oxygen levels), and sound (to monitor snoring) are used in polysomnography, to evaluate sleep disturbances.

Reticular Activating System

Explain the role of the reticular formation in consciousness

The *reticular formation* is a set of interconnected nuclei enmeshed in white fibers that extends from the medulla oblongata to the midbrain. It includes a wide range of centers that play roles in maintaining balance and posture, cardiac and vasomotor centers of the medulla oblongata, pain modulating centers of the raphe nucleus, and sleep-wake cycles.

> Recticular activating system (RAS) wakes up brain; slows at end of day
> Activity maintained by positive feedback from sensory and motor centers

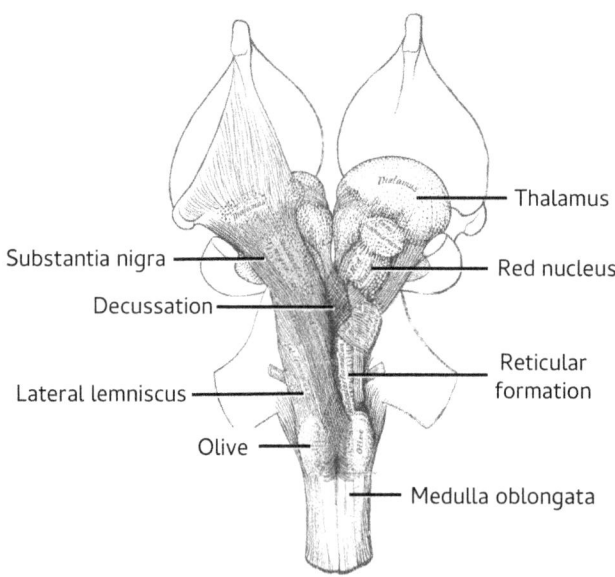

Sensory neurons of the reticular formation project to the thalamus and then to the cerebral cortex, forming the *reticular activating system* (RAS). Input from peripheral senses, proprioceptors, hearing, and vision activate the RAS to wake up the entire brain and focus attention. Thus, pain, body movement, loud sounds, and bright lights cause arousal, keeping us from falling asleep while engaged in activity or responding to threats. RAS activity is maintained by positive feedback from the cerebral cortex and sensory and motor centers. It becomes less responsive toward the end of the day.

Postsynaptic Potentials

Explain why multiple excitatory postsynaptic potentials are required to reach threshold for action potentials

The voltage response of the postsynaptic cell is the *postsynaptic potential* (PSP). At the neuromuscular junction the postsynaptic potential is always a depolarization large enough to initiate an action potential in the muscle fiber. This is the result of Na ion influx through the nicotinic acetylcholine receptor. With metabotropic receptors, the response of the postsynaptic cell may be *excitatory* (depolarization of the membrane that makes it more likely for an action potential to fire) or *inhibitory* (hyperpolarization that makes it less likely for an action potential to be generated). Whether the response is excitatory (EPSP) or inhibitory (IPSP) depends on the receptor and how it is coupled to ion channels in the postsynaptic cell. It is not a property of the neurotransmitter, but the receptor.

EPSP (excitatory postsynaptic potential): depolarization
Multiple EPSPs required to cause action potential in postsynaptic neuron
IPSP causes hyperpolarization, so is inhibitory (makes AP less likely)

At the neuromuscular junction, the postsynaptic potential is always above threshold. This is not the case in the central nervous system, where multiple EPSPs from many synapses must arrive at nearly the same time to cause enough depolarization for an action potential to fire in a postsynaptic neuron. This may be the result of *spatial summation*, where multiple synapses fire at nearly the same time, or *temporal summation*, which is repetitive firing of the same synapse. Usually both occur together.

Neurotransmitters in the Brain

Describe the roles of major neurotransmitters in the brain

The brain uses a wide range of neurotransmitters to control the passage of signals between neurons. For the most part, the role of each neurotransmitter has been sorted out by studying the effects of drugs with known mechanisms on brain activity. Coupled with brain imaging of regions that are active during certain tasks and the ability to trace functional pathways from one region of the brain to another, huge progress has been made in understanding how the most complicated of organs functions.

Drugs can be designed to specifically target a particular receptor and subtype, but targeting a particular region of the brain is difficult. Thus, drugs can have a broad range of effects which causes them to be used in what appear to be unrelated diseases. Agonists act on receptors with the same effects as their natural ligands. Antagonists block the effects of ligands, either by blocking their binding or by binding to allosteric sites that modulate receptor activity.

Five neurotransmitter systems are commonly targeted. We will study each in relation to the drugs that act upon them and the most common indications for those drugs.

GABA (gamma aminobutyric acid) is an inhibitory neurotransmitter. It has two receptor subtypes that are individually targeted. GABA-A receptor agonists are used to treat anxiety and induce anesthesia. Opiates relieve pain by inhibiting transmission by GABA neurons that act on GABA-B receptors in post-synaptic neurons in the brain.

Glutamate is the primary excitatory neurotransmitter in the CNS. In the dorsal horn of the spinal cord, serotonin activates neurons that secrete endogenous opioids. There, opiates inhibit signaling by glutamate from first-order nociceptors to second-order neurons. Glutamate antagonists also induce surgical anesthesia.

Serotonin mediates several effects that influence mood, appetite, and memory. Drugs that increase serotonin levels are antidepressants. Some antidepressants also increase norepinephrine levels.

Norepinephrine is a CNS stimulant. Drugs that increase norepinephrine levels are used to improve focus in attention deficit disorders.

Acetylcholine plays a critical role in memory and attention. It plays a critical role in the synaptic plasticity (ability of synapses to alter their connections) in learning and memory. Drugs to treat Alzheimer's disease increase acetylcholine levels.

Dopamine mediates addictive behavior and is activated by opioids. It plays central roles in executive functions and motor control. Dopamine antagonists are used to treat psychosis. A precursor to dopamine (DOPA) is used to treat Parkinson's disease, a movement disorder.

> Brain imaging identifies functional regions and pathways
> Specific drugs assess roles of neurotransmitters, receptors, transporters

In this chapter, we will study drugs that affect consciousness, anxiety, and pain. A *narcotic* is a psychoactive compound that induces sleep. The term commonly applies to opiates, but has a broader legal definition that includes drugs of abuse. *Analgesics* reduce pain, either at the site of the pain or the transmission and perception of pain in the CNS.

A *sedative* is a CNS depressant (such as alcohol or benzodiazepines). They relieve anxiety (are *anxiolytic*). *Tranquilizer* is another name for a sedative (minor tranquilizers), some of which are antipsychotics (major tranquilizers). A *hypnotic* (or soporific) causes loss of consciousness (sleep). They are commonly called sleeping pills. Often sedatives are used as hypnotics, so are sometimes classified together as *sedative-hypnotics*. Although opiates are narcotics, they are not prescribed for sleep disorders because they do not induce a restful sleep. An anesthetic causes loss of awareness of all sensory input.

Anxiolytics

Explain how sedative-hypnotics induce sleep

Chloral hydrate has been used since the late nineteenth century for treatment of insomnia and as a pre-operative sedative. It is the active component of a "Mickey Finn" which was used by the notorious bartender to render his rich customers unconscious before he robbed them and tossed them out into the alley behind his Chicago bar. Chloral hydrate is metabolized to trichloroethanol, which increases activity of GABA-A receptors.

Barbital, the first of the *barbiturates*, was discovered in 1904 and largely displaced chloral hydrate as a sedative until the ascendance of *benzodiazepines* (*diazepam*) in the 1970s. Barbiturates are now rarely used for insomnia because of their dependence liability and risk of overdose. In high doses, they are used for physician-assisted suicide and in lethal injections (*pentobarbital*). Even so, benzodiazepines are not recommended for treatment of insomnia beyond 2 to 4 weeks for the same reason. They also disturb sleep architecture, decreasing restorative deep slow-wave sleep. *Nonbenzodiazepines* (*zolpidem*) have similar efficacy, but a different chemical structure. They are sometimes called *Z-drugs* because many of their names begin with the letter z.

Common anxiolytics allosterically modulate the activity of *GABA-A receptors*. Barbiturates and benzodiazepines bind to separate sites that are distinct from the binding site of the inhibitory neurotransmitter GABA (*allosteric sites*). GABA-A receptors are ionotropic chloride channels that increase the entry of chloride into post-synaptic neurons. Their activation depolarizes the post-synaptic cell, producing IPSPs and decreasing the likelihood that the postsynaptic cell will fire.

Barbiturates increase the duration of channel opening, which increases GABA efficacy. Each channel is open longer, so has a greater effect. In pharmacology, *efficacy* measures how much of an effect is produced at maximal dose. This explains the high toxicity of barbiturates in over-dose. Barbiturates also block the excitatory neurotransmitter glutamate, so have broader CNS depressant effects than other anxiolytics. Benzo-diazepines increase the frequency of channel opening, which increases GABA potency. *Potency* measures the dose required to produce an effect. A smaller GABA level activates channels more often.

> Anxiolytics increase activity of inhibitory GABA-A receptors
> Efficacy: maximum effect of a drug
> Potency: amount of drug required to cause an effect

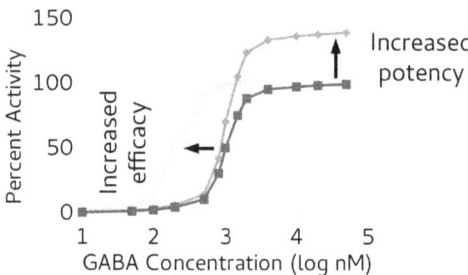

Melatonin, the natural hormone, is an over-the-counter (OTC) agent that can promote sleep. Traditional antihistamines (*diphenhydramine*) and many antidepressants and antipsychotics also have sedating effects. Recently, an inhibitor of orexin receptors (*suvorexant*) has been approved for insomnia. *Orexin* is a neuropeptide that promotes wake-fulness and increases appetite. The primary orexigenic nucleus is in the the lateral hypothalamus.

Primary Afferent Fibers

Distinguish the three primary types of afferent fibers

Afferent neurons (fibers) carry signals from the periphery to the CNS. *Efferent neurons* carry signals from the CNS to effectors (glands and muscle cells). Primary afferent *A-beta (Aβ) fibers* carry non-noxious stimuli (light touch, such as rubbing the skin). They are large diameter and highly myelinated, so transmit signals very rapidly. *A-delta (Aδ) fibers* are poorly myelinated and carry sharp pain signals relatively rapidly. Burning pain is carried by unmyelinated *C fibers*. Since the signals are carried slowly (slow pain). C fiber sensory endings are mul-timodal. They respond to thermal, chemical, and mechanical stimuli.

> Myelinated A-beta fibers carry non-noxious stimuli up the spinal cord
> Lightly myelinated A-delta fibers carry sharp pain signals
> Poorly myelinated C fibers carry slow pain to thalamus

Pain

Describe the mechanism of pain

Pain is a self-reported feeling of severe discomfort (protective mechanism to prevent further injury). *Chronic pain* is pain that lasts a long time (more than 3 months); *acute pain* lasts a shorter time. *Radiating pain* radiates out from a site of injury. *Referred pain* is visceral pain felt some distance from its origin. *Nociceptive pain* is peripheral pain that results from activation of nociceptors in the skin and other organs. *Visceral pain* enters the spinal cord at the same site as a skin sensory receptors (dermatomes). Usually, pain signals come from the skin. So, the brain often interprets pain from a visceral site as pain in the associated skin dermatome, giving the impression that the pain arises from a skin site even though it really arises from an internal organ. *Peripheral neuropathic pain* is the result of damage to or degeneration of the pain fibers (axons) which heightens their sensitivity to stimuli.

> Localized, sharp (fast pain) is carried by myelinated A-delta fibers
> Unmyelinated C fibers carry poorly localized, burning (slow) pain signals
> Both travel up spinal cord through spinothalamic tract to thalamus

Nociceptors are divided into two classes that are associated with two different sensory neuron types. Both fiber types pass up the spinal cord to the thalamus which sends projections to the somatosensory cortex. There pain signal locations are mapped. Rapid responses to injury are sensed by thermal or mechanical nociceptors. This *fast (sharp, first) pain* is localized and of short duration. Fast pain is carried by thin, myelinated (moderately fast conducting) *A-delta (Aδ) fibers*. These fibers are responsible for signaling the withdrawal reflex, when you rapidly remove your hand from a hot object, such as a hot stove.

> TRPV1 is a gated cation channel receptor that is sensitive to a wide
> range of painful stimuli including capsaicin of chili peppers
> Opening the channel initiates an action potential carried by C fibers

Polymodal nociceptors that respond to chemicals as well as extremes of hot and cold and mechanical stimuli are responsible for pain responses that are sensed after the sharp pain has stopped. This *slow (burning, second) pain* is poorly localized and of long duration. It is the throbbing pain you feel after burning your finger. Slow pain is carried by unmyelinated (slow conducting) *C fibers*. This nociceptive pain is responsible for the chronic pain of inflammation and injury.

Ascending Pain Pathways

Describe the pathways that carry pain to the brain

Nociceptors enter the dorsal horn of the spinal cord where they synapse with second-order neurons that cross over (decussate) to the opposite side of the spinal cord before ascending to the thalamus. Thus, pain on the left side is sent to the right side of the brain and vice versa. The dorsal horn synapses are excitatory and use substance P and glutamate as neurotransmitters. Two ascending pathways are used: the *spinothalamic tract* carries pain location information to the somatosensory cortex and the *spinoreticular tract* carries signals that project to multiple regions of the cortex for pain perception. Second-order neurons of both tracts pass through the thalamus before projecting to the cortex.

Functional magnetic resonance imaging (fMRI) of brain activity localizes pain perception to the *cingulate cortex* above the corpus callosum, the *insula* beneath the temporal lobe, and the *prefrontal cortex*. The cingulate gyrus links behavior to motivation, causing us to withdraw from painful stimuli. The insula lights up when we imagine pain or look at images of painful events. The prefrontal cortex is the site of planning and decision-making.

In addition, projections are sent from these pain tracts to lower regions of the brain. The spinothalamic tract sends projections to the *periaqueductal gray* of the midbrain, a nucleus of gray matter synaptic connections surrounding the cerebral aqueduct (thus its name). The ascending second-order neurons of the spinoreticular tract send projections to the reticular formation. Sharp pain keeps you alert.

> Spinothalamic tract signals pain location in somatosensory cortex
> Spinoreticular tract signals to cortical sites perceive stimulus as pain

Analgesia

Explain the mechanism of analgesia in the body

Analgesia is relief of pain. Nociceptors are primarily activated by TRPV1 (*transient receptor potential vanilloid*) cation channels that respond to a wide variety of stimuli, especially capsaicin, which is found in chili peppers. Inflammation causes release of chemicals (inflammatory mediators, such as bradykinin, histamine, leukotrienes and prostaglandins) that sensitize nociceptors to stimuli. Prostaglandins are produced in the periphery by COX-2 (cyclooxygenase-2) enzymes which are inhibited by *NSAIDs* (non-steroidal anti-inflammatories, such as aspirin and ibuprofen). This reduces the release of prostaglandins that contribute to the pain of inflammation at the level of the nociceptor response.

Pain may also be controlled at the level of the spinal cord where C fibers enter the dorsal roots at the same level as sensory neurons for cutaneous mechanoreceptors which are large, myelinated A-beta (Aβ) fibers. These are distinct from the smaller, myelinated Aδ fibers that carry fast pain to the spinal cord. When signals from C fibers enter the spinal cord, they synapse with second-order neurons that pass across the gray commissure, transmitting pain signals to the cerebral cortex.

If signals from cutaneous mechanoreceptors arrive at the same time, those signals inhibit the passage of pain signals from the active C fibers. Thus, rubbing a "hurty" makes it feel better. The mechanism is called the *gate control theory*, since signals from Aβ fibers control opening of the gate for C fiber signals to be passed on through the spinal cord to be perceived by the brain as pain. Strong pain signals can overcome inhibition by cutaneous mechanoreceptor signals, so rubbing does not always work.

> Pain can be relieved by activating cutaneous mechanoreceptors
> Signals from A-beta fibers inhibit pain transmission by C fibers
> to second-order neurons in the spinal cord (gate control theory)

Descending Pain Pathways

Describe the descending pathways that influence transmission of pain

Pathways from the brain also influence transmission of signals from pain receptors. This pathway originates in the *periaqueductal gray* (PAG) of the midbrain which sends inhibitory GABA interneurons to the *raphe nuclei* in the medulla. The GABA receptors in this pathway are metabotropic G-protein coupled receptors (*GABA-B receptors*), which are distinct from the GABA-A receptors that mediate anxiety. When these neurons are active, they prevent the raphe nuclei from sending inhibitory signals to the dorsal horn where nociceptors enter the spinal cord. Inhibition of the descending pathway allows transmission of pain signals up the ascending pathway to the brain.

Endogenous opioids are small peptides that bind to opioid receptors. The three receptor classes are named according to how they were discovered. Morphine binds to *mu receptors*; the drug ketocyclazocine binds to kappa receptors; *delta receptors* were first isolated from murine vas deferens tissue. *Endorphins* bind primarily to mu receptors. Enkephalins bind to mu and delta receptors. Dynorphins prefer kappa receptors. The primary target for analgesics is the mu receptors.

Endogenous opioid peptides are released in stressful situations. They are inhibitory. For example, athletes can often continue to play after being injured. Stimulation of the PAG or raphe nuclei causes analgesia, without interfering with touch and temperature sensations. Presynaptically, opioids block voltage-gated channels to reduce neurotransmitter release. Opioids inhibit ascending signals to the thalamus, blocking transmission of pain sensations to higher centers in the brain.

At the same time, opioids block inhibitory signals sent from the periaqueductal gray to the raphe nucleus. This activates the descending pain pathway to the dorsal horns where nociceptors synapse with second-order neurons. In the dorsal horn, serotonergic neurons from the raphe nucleus secrete endogenous opioid peptides which bind to *opioid receptors* and block transmission of pain signals up the spinal cord.

> Serotonergic neurons from raphe nucleus descend to dorsal horns
> Serotonin activates inhibitory neurons that release endogenous opioids
> Opioids block pain signals from nociceptors to second-order neurons

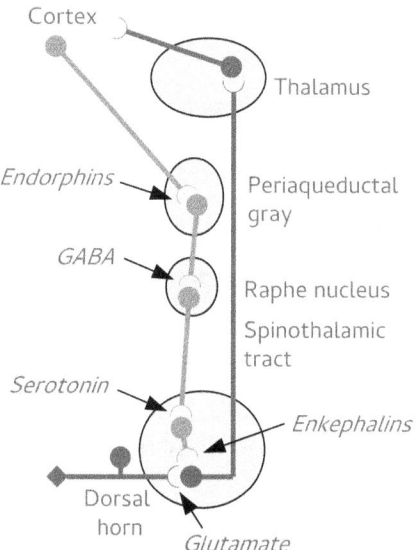

In the dorsal horn, opioids inhibit neurotransmitter release by nociceptors and postsynaptically open K ion channels that hyperpolarize the membrane. This decreases the likelihood an action potential will fire in the ascending second-order neuron. The end result is reduced transmission of pain signals through the ascending pathway.

Opioids

Describe the effects of opioids

Opiate refers to a drug derived from opium, which includes *morphine* and *codeine*. *Heroin* (diacetylmorphine) is a semisynthetic drug derived from morphine. Codeine and heroin are converted to morphine by liver enzymes. When injected, heroin rapidly passes through the blood-brain barrier, causing euphoria. Opioid is a broader term. It refers to any drug (*meperidine, oxycodone*) or endogenous peptide that binds to opioid receptors to produce a opiate effect.

Opioid receptors not only modulate pain pathways. They are also found in the gastrointestinal tract where they are responsible for maintaining gut motility. Inhibition by opiates reduces peristalsis to relieve diarrhea, but may then cause constipation. Opioids (particularly codeine) are cough suppressants. In addition, decreased activity of inhibitory GABA neurons by opiates increases dopamine release, causing activation of the reward pathway and feelings of pleasure that can be addictive. Opioids reduces sensitivity to carbon dioxide chemoreceptors, inhibiting respiratory response to carbon dioxide. Respiratory depression is dose-dependent and the most common cause of death due to overdose. *Naloxone* is an opioid antagonist that blocks opioid binding and reverses respiratory depression due to opioid overdose.

Anesthesia

Describe the stages and drugs used in anesthesia

An analgesic relieves pain without loss of consciousness. Anesthetics depress the CNS causing the loss of consciousness of all sensory modalities, without loss of vital functions. *Local (regional) anesthetics* (lidocaine) are basic drugs that block voltage-gated Na channels in axons. They reduce action potential firing, especially in rapidly firing neurons to stop transmission of sensory information from the periphery to the CNS. The unprotonated base diffuses across membranes where it is ionized and trapped inside cell to block open voltage-gated Na channels on the inner surface of the axonal membrane.

Conscious sedation couples moderate sedation with analgesia. The patient stays conscious and cardiorespiratory function is normal. Conscious sedation is used in minor surgical procedures, such as dentistry and cosmetic surgery. For example, a benzodiazapine anxiolytic (midazolam) is often combined with a strong opioid analgesic (fentanyl). Ketofol is a 1:1 mixture of ketamine (analgesic) with propofol (sedative and amnesic).

Ketamine is an *NMDA receptor antagonist* that induces dissociative anesthesia: catalepsy (muscular rigidity), amnesia, and analgesia. In addition, it increases blood pressure and cardiac output through sympathetic activation. *NMDA receptors* (N-methyl-D-aspartate) are one of three subclasses of ionotropic glutamate receptor. Binding of glutamate and glycine opens channels that allow cation entry, depolarization, and excitation of postsynaptic neurons. Ketamine blocks the channel to depress excitatory signals in the CNS. The same NMDA receptor channel is also blocked by the gaseous anesthetic nitrous oxide. Propofol blocks GABA-A receptors and sodium channels. It has largely replaced thiopental (a barbiturate) for induction and maintenance of anesthesia.

General (surgical) anesthesia is reversible CNS depression that results in loss of sensation and response to and perception of external stimuli. Early anesthetics include ether and chloroform. Modern surgical anesthetics are volatile halogenated hydrocarbon liquids (*isoflurane*) that are administered with oxygen by a recirculating system that removes carbon dioxide to reduce cost and waste. They are not flammable or explosive. Depth of anesthesia can be rapidly altered by changing the inhaled concentration of drug.

Isoflurane is a GABA and glutamate receptor agonist. It increases the sensitivity of GABA-A receptors to GABA inhibition of neurotransmission in somatosensory pathways. It has a narrow therapeutic window. Administration must be closely monitored to prevent progression to medullary paralysis. Long-acting neuromuscular blockers (*succinylcholine*) are required for thoracic and abdominal surgeries.

Review Questions

1. What are the major divisions of the nervous system?
2. What are the six divisions of the brain? Which are found in the brainstem?
3. What are the two main divisions of the diencephalon and their functions?
4. What is the function of the cerebrum?
5. What are the overall functions of the ANS and its two divisions?
6. What is the function of astrocytes?
7. What are the three layers of the meninges in brain and spinal cord?
8. Describe the production, flow, and function of cerebrospinal fluid (CSF). Where is the dural sac?

9. How do nuclei differ from tracts? What are the major types of tracts in the brain?

10. How are the two hemispheres of the cerebrum connected?

11. What is an EEG and how is an EEG recording divided into distinct components?

12. What can imaging methods tell us about the brain?

13. How are circadian rhythms established?

14. What are the stages of sleep? What is the RAS?

15. What is the difference between an EPSP and an IPSP?

16. What is the mechanism of action of anxiolytics? What kind of receptor is GABA-A (ionotropic or metabotropic)?

17. How do the three primary afferent fibers differ?

18. Distinguish between slow and fast pain. What is the gate control model of pain inhibition?

19. What are the components of the ascending pain pathway?

20. What is the mechanism of action of most analgesics?

21. How do descending pain pathways inhibit pain transmission? What neurotransmitters are involved?

22. How do opiates act to inhibit transmission of pain?

23. What receptors mediate local, conscious, and general anesthesia?

Somatic Motor and Executive Functions

The *brain* must make sense of sensory input (*sensations*) coming from all over the body. Sensory input from body organs (other than pain) is almost entirely integrated within the brainstem. Input from cutaneous receptors, proprioceptors, and special senses are integrated at higher levels in the brain. *Cutaneous receptors* map sensory input to the homunculus of the post central gyrus (*somatosensory cortex*). *Proprioceptors* (including equilibrium receptors of the inner ear) send information primarily to the *cerebellum*, which is responsible for automatic control of movement. *Special senses* send fibers to special areas of the brain that are named for the sense: olfactory cortex, visual cortex, gustatory cortex, and auditory cortex.

Nearby cortical areas (*association areas*) interpret the raw sensory information that is sent to the *primary areas*. Interpretation of sensations as *perceptions* involves an integration of the new sensory information with stored memories. For example, I recognize an object on the table as a coffee mug because of past associations of similar visual input from other mug-shaped object that I have used or seen other use to drink coffee. The light rays that impinge on the retina and are sent to the visual cortex are pure sensations. Recognizing that the object is a coffee cup is a perception.

Normally, our perceptions are accurate and we act upon them. If I see a dark colored liquid in the cup and know that you have poured it out for me from the coffee maker, I am likely to taste it. I will be a bit cautious because past experience tells me that the liquid may be rather hot. After I taste it, I can confirm whether it is good coffee or some less desirable dark liquid. If it is good, I drink and enjoy it. Thus, sensation is turned into perception, we act on the perception, and our action confirms (or does not confirm) our perception.

This whole system can get messed up at any point. Sensation itself may fail us. A finger that has been severely burned can no longer be used to judge the texture of an object because its touch receptors have been damaged. If perception fails, the problem can be serious. If this failure of perception threatens our well-being or that of others, it is classified as a *psychosis* and requires treatment. The *motor system* may fail. We might shake when we make a movement or be unable to make a coordinated movement. In this lecture, we will look into all of these topics: how they should work, how they might fail, and what can be done about it.

Limbic System

List the components of the limbic system and their functions

The *limbic system* regulates motivation and emotion and consolidates memories. It is at the border between higher and lower centers, where subcortical centers meet the cerebral cortex. The limbic system ensures propagation of the species by motivating parenting and feeding functions and preserving emotional memories.

The major structures of the limbic system are the cingulate gyrus, hippocampus, fornix, and amygdala. The *cingulate gyrus*, which lies directly above the corpus callosum, regulates emotional memory and learning. The *hippocampus* stores and retrieves new long-term memories. The *fornix* connects the hippocampus with the hypothalamus. The *amygdala* (amygdaloid body) links emotions with memories, so has been called the fear center.

> Limbic system consists of cingulate gyrus: emotional memory and learning
> Hippocampus: consolidates new memories
> Amygdala links emotions with memories (fear center)

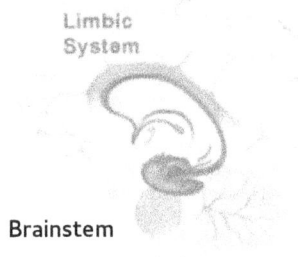

Limbic
System

Brainstem

Spinal cord

Cerebrum

Describe the organization of the cerebrum

The *cerebrum* is highly folded with prominent infoldings (*sulci*) and ridges (*gyri*). The central sulcus divides primary motor (precentral) from sensory (postcentral) regions. *Cerebral lobes* correspond to the names of their overlying bones: frontal, parietal, temporal, and occipital. Functions are integrated in a thin layer of gray matter on the surface of the cerebrum (the *cerebral cortex*).

> Cerebrum processes sensations and plans and initiates movements
> Cerebrum is subdivided into lobes named according to cranial bones
> Cerebral cortex is thin outer layer of cerebrum

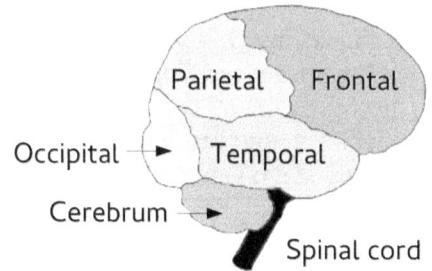

Primary areas in the cortex are involved in sensation or execution of movement. The *primary sensory cortex* in the postcentral gyrus receives somatic sensory information. The sensory cortex is mapped as a little person (*homunculus*) in proportion to sensory sensitivity. Olfaction, vision, gustation (taste) and audition (hearing) have distinct primary sensory areas. The precentral gyrus directs voluntary movements by sending out upper motor neurons, so is called the *primary motor cortex*.

Association areas integrate multiple inputs to generate meaningful patterns. They are adjacent to primary areas and include the premotor area and somatosensory, visual, and auditory association areas. The *premotor area* integrates input from sensory association areas, basal ganglia (nuclei at the base of the cerebrum), and thalamus to direct output of upper motor neurons from the primary motor cortex.

Integrative centers integrate information from association areas and direct complex motor activities. Wernicke's area interprets language. Broca's area produces speech. The *prefrontal cortex* is the seat of judgment, reasoning, and planning.

> Primary areas of cerebral cortex receive inputs and send out commands
> Association areas bring together multiple inputs to generate patterns
> Integrative areas integrate information from many parts of the brain

Mood Disorders

Describe the mechanism of action of antidepressants

Depression refers to changes in emotional responses that impair normal work, family, or social function. To be diagnosed with *major depressive disorder*, the symptoms must be present for at least two weeks. Antidepressants are drugs that treat depression by elevating mood. Most antidepressants elevate serotonin in brain synapses. This observation has lead to the biogenic amine hypothesis of depression: low levels of monoamines (catecholamines) cause depression.

One of the first successful drugs in this class (*phenylzine*) inhibits MAO (*monoamine oxidase*), a mitochondrial enzyme that degrades catecholamines. *MAO inhibitors* decrease catabolism of *catecholamine neurotransmitters* (dopamine, NE, serotonin) that have been taken back from the synaptic cleft into axon terminals. This increases catecholamine signaling.

Dopamine is the major neurotransmitter in reward pathways and is largely responsible for the addictive side effects of opiates. However, the success of selective *serotonin reuptake inhibitors* (SSRIs, such as *fluoxetine*) in treating depression strongly suggests that the antidepressant activity of MAO inhibitors is mediated primarily by increased levels of *serotonin*. SSRIs specifically block the serotonin reuptake transporter (SERT) in axon terminals.

Tricyclic antidepressants (TCAs, such as *imipramine*) inhibit reuptake of serotonin and norepinephrine from the synaptic cleft back into axon terminals by SERT (serotonin reuptake transporter) and NET (norepinephrine transporter). *Serotonin-norepinephrine reuptake inhibitors* (SNRIs, *duloxetine*) also decrease serotonin and norepinephrine reuptake and have better side-effect profiles. Neither drug class has significant effects on dopamine reuptake.

Ecstasy (*MDMA*, methylenedioxymethamphetamine), like other amphetamines, competes with catecholamines for reuptake by monoamine transporters and inhibits *vesicular monoamine transporters* (VMATs). Inhibition of VMATs causes accumulation of monoamines in the cytosol of the axon terminal. At the same time, ecstasy causes phosphorylation of reuptake transporters, which reverse direction to cause neurotransmitter efflux through the reuptake transporters. The effects of MDMA are primarily on the serotonin system. Thus, the serotonin levels increase and the elation experienced with ecstasy is consistent with the serotonin hypothesis of depression.

Cell bodies of neurons that secrete serotonin in the brain are clustered in the raphe nucleus. Serotonin is released from neurons that project to the limbic system (elevates mood), hypothalamus (suppresses appetite), hippocampus (memory consolidation), and the frontal cortex (improved cognition). These beneficial effects make increasing serotonin levels a popular goal, but doing so with SSRIs and related drugs is problematic. Natural methods of increasing serotonin levels include exposure to light and exercise, particularly outdoor exercise. Tryptophan is a precursor to serotonin and can be supplemented in the diet by wise food choices (salmon, tofu, cashews, beans, eggs, whole grains). In addition, vitamin B6 is required for synthesis of serotonin and related neurotransmitters.

Bipolar disorder is associated with mood swings (episodes of depression alternating with mania). Characteristic of *mania* is high psychomotor activity and irritability. *Mood stabilizers* moderate extremes of moods. *Lithium carbonate* interferes with several signaling pathways that influence mood. *Antiseizure drugs* are also effective in manic episodes bipolar disorder. *Carbamazepine* blocks sodium channels in axons, decreasing the rate of action potential firing. *Valproate* inhibits neuronal signaling by activating GABA pathways.

Common antidepressants elevate serotonin levels by blocking reuptake
Many antidepressants also inhibit norepinephrine reuptake by axon terminals

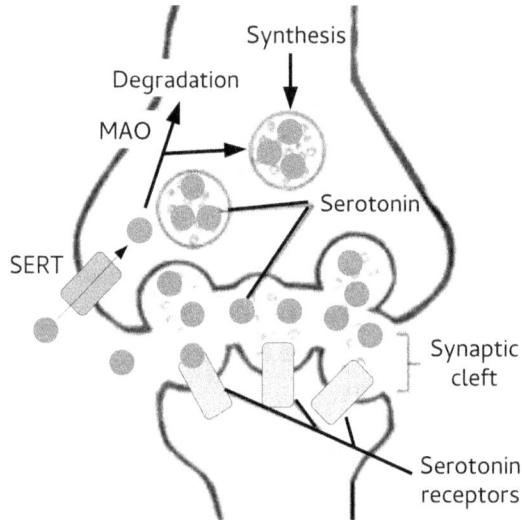

Learning and Memory

Explain the concepts of association in memory formation and consolidation

Memory (retention of what is experienced) is encoded in synaptic connections. *Declarative memories* are those of events, experiences, and ideas (facts). New declarative memories are first stored in the hippocampus before being more broadly distributed to the cerebral cortex. *Procedural memories* are learned motor behaviors (skills). Memory for the automatic part of a complex movement is encoded in the cerebellar cortex. *Short-term memory* holds memories in storage for a short period of time. It is distinct from *working memory* that is used to process and use information. *Long-term memory* stores large amounts of information for long periods, but is subject to fading (loss of the memory) if it is not rehearsed from time to time.

The mechanisms underlying memory formation in hippocampus and cerebellar cortex are similar. *Synaptic plasticity* is the ability of a synapse to be remodeled in response to repeated stimuli or new experiences. Repetition and association strengthen synaptic connections. Disuse weakens them or causes synaptic connections to atrophy. The *classical conditioning* experiments of Ivan Pavlov are an example of associative learning. Pavlov caused dogs to salivate upon ringing of a bell by presenting food again and again whenever the bell was struck. The dogs associated food with the bell. We do the same when we sequence a series of events, integrate new memories into old ones, or use multiple senses to take in information. It is easier to remember a series of steps than individual events that are unrelated. We make sense of new information more readily if it is integrated into knowledge we already have.

Repeated stimulation of a synaptic connection strengthens it (*Hebb's rule*: cells that fire together, wire together). Synapses can be strengthened by increasing the amount of neurotransmitter released or the number of receptors in the post-synaptic membrane. In *long-term potentiation* (LTP), a train of high-frequency impulses across a synapse leads to long-term changes in synaptic strength. The late phase of LTP results from increased gene expression, protein synthesis, and formation of new synapses.

> Learning strengthens some synapses and weakens others
> Long-term potentiation: repeated stimulation of synapse increases
> neurotransmitter release and postsynaptic receptor number

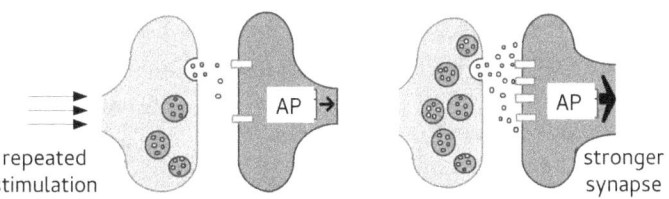

Memory consolidation converts short-term to long-term memories and occurs in two stages. *Synaptic consolidation* is the strengthening of synaptic connections and is identical with late-phase LTP. It is confined to the hippocampus and occurs over a period of hours. *Systems consol-*

idation distributes memories more widely over the cerebral cortex. These memories are organized as an *engram* with multiple complex synaptic connections. Stabilization of memories in the cortex takes days to years, as new memories are integrated with existing memories.

In the standard model of systems consolidation, new memories undergo LTP in the hippocampus and are only weakly mapped to the cerebral cortex. As new memories are rehearsed and integrated into existing memories, the synaptic connections for the new memory in the cerebral cortex are strengthened. As it does so, the new memory becomes less dependent on the hippocampus. With repetition and association, the memory is mapped in cortical connections to become a long-term memory that does not depend on the hippocampus.

Memory formation begins in hippocampus and is weak in cerebral cortex
With repetition and association cortical memory strengthens
by systems consolidation to become independent of hippocampus

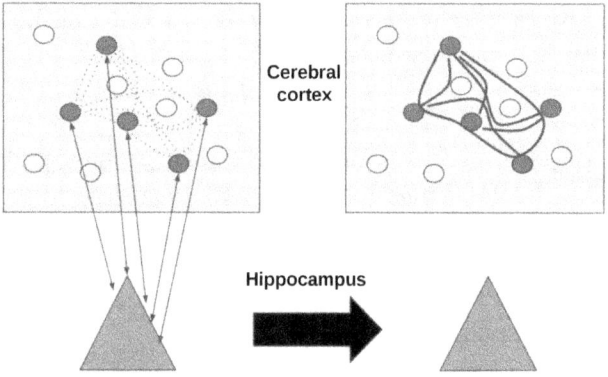

Memory Neurons

Describe the roles of cholinergic neurons and NMDA receptors in memory formation

The brain has both muscarinic and nicotinic cholinergic receptors that respond to the neurotransmitter acetylcholine. *Scopolamine* is an antagonist at muscarinic cholinergic synapses (*antimuscarinic*). It is used to treat motion sickness and post-operative nausea and vomiting. Scopolamine prevents signaling from the vestibule (equilibrium center in inner ear) to the vomiting center in the brain. It can also cause severe memory loss. On the other hand, drugs which activate cholinergic receptors (such as nicotine) enhance cognition and alertness.

Most cholinergic neurons arise in the *basal forebrain* which lies next to the diencephalon at the base of the frontal lobe. These neurons project to the cerebral cortex (attention), hippocampus (memory encoding), and amygdala (emotional memory). Cholinergic signaling is critical to hippocampal synaptic plasticity in learning and memory encoding. It facilitates long-term potentiation (LTP) in both hippocampus and amygdala. The basal forebrain is damaged in early Alzheimer's disease, consistent with the hypothesis that acetylcholine plays a central role in its characteristic short-term memory loss (the inability to convert new information into memories). Furthermore, memory impairment in Alzheimer's disease is often associated with attention deficits.

The *NMDA receptor* is an ionotropic glutamate receptor that is involved in long-term potentiation (LTP). NMDA receptors detect synaptic coincidence, forming stronger connections when two neurons are repeatedly stimulated together. NMDA receptor systems play a central role in learning and memory. NMDA receptors mediate memory encoding in the hippocampus and amygdala but are not required for maintenance of long-term memories.

> Cholinergic neurons arise in the basal forebrain and project to cerebral cortex (attention), hippocampus (memory encoding), amygdala (emotional memory)
> NMDA receptors mediate LTP learning and memory by synaptic coincidence

Alzheimer's Disease

Explain the basic mechanisms of memory loss in Alzheimer's disease

Alzheimer's disease causes the loss of neurons in the frontal lobe of the cerebral cortex and in the hippocampus, sites that process and maintain memories. Neurodegeneration leads to formation of distinctive *amyloid plaques* (extracellular deposits) and *neurofibrillary tangles* (aggregates of tau proteins). It leads to memory loss, confusion, cognitive loss, and mood disorders. The most common early symptom is short-term memory loss which causes disorientation, mood swings, and social withdrawal. It is the most prevalent degenerative CNS disorder (70% of dementias).

Symptoms of Alzheimer's disease are improved with cholinesterase inhibitors that can cross the blood brain barrier. *Acetylcholinesterase inhibitors* (*donepezil*) decrease the rate of degradation of acetylcholine in synaptic clefts, which increases cholinergic signaling in the brain. The result is a slower progression of short-term memory loss, especially in patients with early Alzheimer's disease.

An alternative approach is to use *NMDA receptor antagonists* (*memantine*). This should strike you as counterintuitive given what has been said about the role of NMDA receptors in memory formation. However, overexcitation of the NMDA system causes excitotoxicity. Excess stimulation by glutamate increases intracellular Ca and this high cellular Ca activates enzymes that cause programmed cell death (*apoptosis*). NMDA antagonists minimize excitotoxic damage.

> Alzheimer's disease is characterized by memory loss with amyloid plaques
> Acetylcholinesterase inhibitors improve memory and attention deficits
> Memantine blocks NMDA receptor-induced neurotoxicity

Attention Deficits

Attention deficit hyperactivity disorder (ADHD) is a neurodevelopmental disorder characterized by poor impulse control, hyperactivity, and short attention span. ADHD is commonly treated with CNS stimulants (*methylphenidate, amphetamine*) that heighten awareness and improve focus by blocking norepinephrine and dopamine reuptake into presynaptic neurons. Amphetamines also increase dopamine release by axon terminals. These drugs activate circuits in the prefrontal cortex that maintain awareness and regulate behavioral control, attention, and motivation.

An alternative approach is to use non-stimulants, which have lower abuse potential and also lower success rates than stimulants. *Atomoxetine* is more specific for norepinephrine reuptake and has less effect on the dopaminergic system. *Clonidine* is an alpha-2 adrenergic agonist. In the prefrontal cortex, it binds to postsynaptic alpha-2 receptors. Both increase norepinephrine signaling in the prefrontal cortex. In the vasomotor center it binds to presynaptic alpha-2 receptors, reducing sympathetic effects and lowering blood pressure.

> Attention and impulse control are improved by increasing NE and dopamine

Executive Functions

Describe the mechanisms involved in judgment and decision-making

Executive functions are cognitive processes that control behavior. To carry out an intended behavior requires attention, planning, reasoning, problem solving, and control of impulses. Many of these systems are mediated by dopaminergic neurons. Focus and attention are improved by dopamine reuptake inhibitors, such as those used for ADHD. The symptoms of erratic behaviors in psychosis are diminished by dopamine antagonists, consistent with an overactive dopamine system in the prefrontal cortex. Further support for this hypothesis comes from drugs (cocaine, amphetamines) that increase dopamine levels and can trigger psychotic symptoms. Dopamine is also an important neurotransmitter for motor control. Low dopamine levels cause movement disorders with Parkinson's-like symptoms.

Three *dopaminergic pathways* that connect to GABAergic neurons mediate these effects. Two arise from dopaminergic neurons in the *ventral tegmentum* (VTA) in the midbrain, next to the cerebral aqueduct. The *mesocortical pathway* connects the ventral tegmentum to the prefrontal cortex. This pathway is associated with cognitive control, motivation, and emotion. The *mesolimbic pathway* connects the ventral tegmentum to nuclei in the ventral striatum, which lies between the basal forebrain and hypothalamus. This region is important in reward and motivation circuits. It is largely responsible for addiction. The *nigrostriatal pathway* connects the substantia nigra with the dorsal striatum. It is responsible for ensuring smooth movements.

> Dopamine plays a central role in executive functions that require judgment
> Mesocorticolimbic pathways regulate cognition, motivation, and emotion
> Nigrostriatal pathways regulate and smooth movements

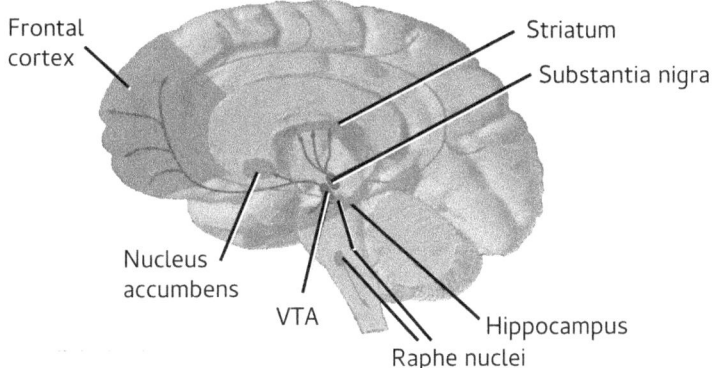

Frontal cortex · Striatum · Substantia nigra · Nucleus accumbens · VTA · Hippocampus · Raphe nuclei

Psychosis

Describe psychosis and the drugs used to treat it

Psychosis is loss of contact with reality that impairs daily living. It may involve delusions (strong beliefs in something not based in reality), hallucinations (seeing, hearing, or feeling something not there), illusions (distorted or misleading perceptions), disorganized behavior (highly inappropriate actions), and paranoia (irrational distrust of people and objects). The most common psychosis is *schizophrenia Positive symptoms* add to normal behavior (such as hallucinations or nonsense speech). *Negative symptoms* subtract from normal behavior (such as inability to feel pleasure or express emotions).

Neuroleptics are antipsychotic drugs. Most are dopamine D2 receptor antagonists in the frontal cortex. They reduce positive feedback of neural circuits that project back to dopamine neurons, producing a calming (tranquilizer) effect, but are not sedatives (do not put the patient to sleep). *Typical antipsychotics* (conventional antipsychotics) have been available since the 1950s. They are through to reduce symptoms of schizophrenia through binding to D2 receptors. Typical antipsychotics include phenothiazines (*chlorpromazine*) and non-phenothiazines (*haloperidol*). *Extrapyramidal symptoms* (EPS), Parkinson's-like movement disorders, are common. These are due to dopamine antagonism in the striatum, which mimics the loss of dopaminergic neurons in Parkinson's disease. Sustained muscle spasms (*acute dystonia*) may progress to *tardive dyskinesia* (involuntary, repetitive, slow body movements) which are irreversible.

Atypical antipsychotics (*risperidone*) are D2 antagonists with greater affinity for serotonin receptors than dopamine receptors. They may be combined with lithium and valproic acid to treat bipolar disorder. They have fewer extrapyramidal symptoms but often cause weight gain. *Aripiprazole* has a more complex dopamine receptor profile, described as functional selectivity. Like other atypical antipsychotics, it also has high affinity for serotonin receptors. Unlike earlier drugs, it is a partial agonist at some D2 receptors. Such "third-generation" antipsychotics cause less weight gain than earlier atypical antipsychotics.

> Most antipsychotics are antagonists of D2 dopamine receptors
> Interference with dopamine signaling may cause Parkinson's like symptoms
> Atypical antipsychotics inhibit serotonin more than dopamine signaling

Motor Pathways

Describe the pathways from cerebral cortex to skeletal muscles

The motor homunculus in the *primary motor cortex* is a cortical map for motor signals. *Upper motor neurons* arise from the primary motor cortex in the precentral gyrus. Their activity is influenced by input from the premotor cortex, basal ganglia (nuclei), and cerebellum. The *premotor cortex* is a frontal lobe association cortex that coordinates learned movements. The *substantia nigra* sends inhibitory signals (dopamine) to the *basal ganglia* to suppress unwanted movements. The basal ganglia (nuclei) includes the *striatum*, with its two divisions (*caudate* and *putamen*) and the *globus pallidus*. They modify motor neuron signals to ensure that movements are made smoothly. The *cerebellum* controls learned movements, correcting output of the upper motor neurons to match the intended movement with actual movement.

Basal ganglia: striatum (caudate and putamen) and globus pallidus modify motor neuron output to ensure movements are made smoothly

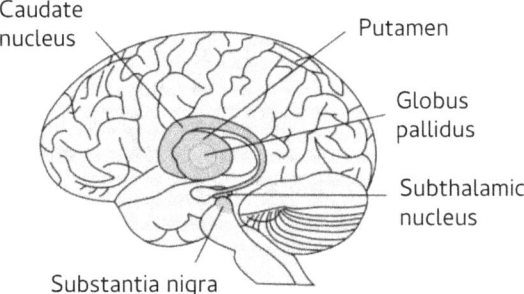

Most motor neurons pass through *medullary pyramids* and decussate (pass to the opposite side of the body) through the corticospinal (pyramidal) pathway. *Lower motor neurons* innervate muscles through cranial and spinal nerves via a single *motor unit*, which consists of the lower motor neuron and all the muscle fibers it innervates. Lower motor neuron cell bodies are in the brainstem or spinal cord.

Upper motor neurons arise in primary motor cortex, decussate in pyramids and send out lower motor neurons to skeletal muscles

Cerebellum

Describe the role of the cerebellum in controlling body movements

The *cerebellum* receives input from proprioceptors and visual and vestibular (balance) centers. The *spinocerebellar tract* carries signals from proprioceptors to the cerebellum. Purkinje cells in the cerebellar cortex have extensive synaptic connections. Motor signals from the primary motor cortex and basal ganglia enter the cerebellum from the pontine nucleus of the pons. At the same time the cerebellum directly receives input from joint and muscle proprioceptors, vestibular apparatus of the inner ear, and the eyes.

Cerebellum monitors output of proprioceptors, vestibular apparatus, and eyes to judge body position and compares it with intentions
Corrections to motor error modify output of upper motor neurons

Together, these inputs monitor body position and modulate activity of upper motor neurons to ensure smooth and coordinated movements. The cerebellum compares the difference between an intended and actual movement (*motor error*). It reduces motor error by feedback and learns complex movements through practice.

Epilepsy

Describe the mechanism of action of drugs used for epilepsy

Epilepsy is a neuromuscular disorder characterized by chronic, recurring seizures due to hyperexcitability of CNS neurons. It may cause muscle twitching or alterations in consciousness (confusion, blank staring, repetitive blinking). Antiepileptics (*anticonvulsants*) minimize seizure activity by suppressing neuronal activity. *Status epilepticus* is continuous seizure activity that may lead to coma or death. It is a medical emergency. First-line drugs are benzodiazepines (diazepam or lorazepam). If these fail phenytoin or phenobarbital are used.

Benzodiazepines increase inhibitory GABA binding and the frequency of Cl channel opening. This hyperpolarizes the resting membrane potential of postsynaptic cells and suppresses neuronal transmission. *Phenytoin* binds to the inactive form of voltage-gated Na channels in the motor cortex. This reduces repetitive firing of action potentials and minimizes spread of disruptive electrical signals. Phenobarbital prolongs and potentates the effects of inhibitory GABA signaling. Valproic acid combines these effects. It blocks voltage-gated Na and Ca channels and increases GABA inhibition.

> Anticonvulsants block Na channels and increase inhibitory GABA

Parkinson's Disease

Describe the pathology and treatment of Parkinson's disease

In *Parkinson's disease* dopamine secreting neurons in the substantia nigra degenerate. These neurons normally secrete dopamine to inhibit excitatory cholinergic neurons of striatum in order to relax antagonist muscles during movement. Low dopamine levels cause tremors, muscle rigidity, bradykinesia (slow movement), and altered posture. Parkinson's treatment elevates dopamine and decreases acetylcholine to bring inhibition and excitation into balance.

Dopamine does not cross blood-brain barrier, so its precursor (L-DOPA = *levodopa*) is administered. Levodopa is converted to dopamine in the CNS. *Carbidopa* is a DOPA decarboxylase inhibitor that reduces peripheral conversion of levodopa to dopamine. *Entacapone* is a COMT inhibitor that stops conversion of levodopa to COMT in the periphery. It also increases DOPA delivery to CNS. Stalevo is a proprietary drug that combines levodopa, carbidopa, and entacopone. *Anticholinergic drugs* (*benztropine*) block excess excitatory cholinergic stimulation of motor neurons in corpus striatum. Benztropine is also used to treat

Parkinson's-like movement disorders (extrapyramidal symptoms) due to antipsychotics.

> Parkinson's disease is treated by increasing dopamine levels
> Extrapyramidal symptoms of antipsychotics are treated with anticholinergics

Language and Speech

Describe the pathways for understanding and producing speech

Speech perception is the process by which words that are read or heard are interpreted and understood largely by the left hemisphere of the cerebral cortex. Simply looking at words activates the visual cortex of the occipital lobe. Hearing words is somewhat more complex and involves the primary auditory cortex and nearby regions of the temporal lobe. These include the *paralimbic cortex* adjacent to and within the limbic system, which provides access to long-term and emotional memory centers, and the superior temporal region that gives meaning to words that are heard.

Wernicke's area (general interpretive center) is within the superior temporal gyrus between the auditory cortex and inferior parietal lobule. This area interprets phonetic sequences (which represent the sounds of words) whether they are heard, recalled from memory, or generated as new speech. Semantic processing (recognizing words in context) is primarily the role of the nearby *angular gyrus.* Simply pronouncing words (such as the days of the week in order) primarily activates the *supramarginal gyrus,* which arches over the lateral sulcus. The angular gyrus and supramarginal gyrus also receive input from the visual cortex.

> Primary auditory cortex maps sound frequencies from inner ear
> Wernicke's area recognizes words which angular gyrus puts into context
> Broca's area directs motor cortex to produce sounds

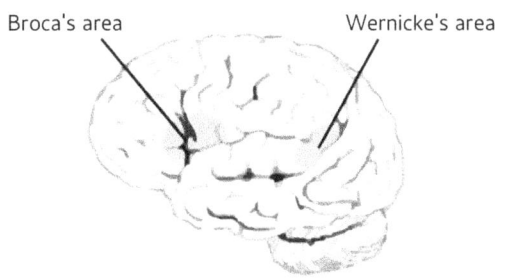

Broca's area   Wernicke's area

Speech production translates thoughts into audible speech. *Broca's area,* (speech center) lies just anterior to the premotor cortex. It coordinates motor cortex output to the muscles of mouth and tongue that produce speech. It is most active, along with Wernicke's area, when generating words.

Damage to Wernicke's area usually causes *receptive aphasia*, in which speech is fluent but nonsensical. Sufferers might replace the desired word with one that sounds correct, but is not. They are often unaware that what they said is not what they intended. Damage to Wernicke's also compromises comprehension.

Damage to Broca's area causes difficulty in speech production (*expressive aphasia*). Speech is not fluid, but labored and difficult. Sentences are simplified to telegraph speech, such as saying "drive store mom" instead of "my mom drove me to the store today." Comprehension is not compromised. The sufferer is aware of their difficulty. Aphasia of both types is a common consequence of stroke.

> Listening activates auditory cortex and nearby temporal centers
> Pronouncing words activates premotor and nearby motor cortex
> Generating words activates Broca's and Wernicke's areas

Review Questions

1. What are the structures of the limbic system and their functions?
2. Distinguish primary, associative, and integrative regions of the cerebrum.
3. What is the most common neurotransmitter deficiency in depression? How is it treated?
4. How are memories consolidated and preserved?
5. What is the mechanism of long-term potentiation? What role do NMDA receptors play in LTP?
6. What neurotransmitter deficiency is most likely the cause of memory loss in dementia? How is Alzheimer's disease treated?
7. What neurotransmitters are primarily involved in attention and impulse control? How is ADHD usually treated?
8. What are the three dopaminergic pathways that regulate executive functions?
9. On what receptor system do typical antipsychotics work? What are extra-pyramidal symptoms?
10. Describe the motor pathways from the motor cortex to skeletal muscles.
11. What are the roles of the basal ganglia (nuclei) and cerebellum in controlling movement?

12. How is epilepsy treated? What is the primary failure in Parkinson's disease?

13. What are the roles of Wernicke's and Broca's areas in speech?

Figure Credits

Cell Biology

Phospholipid bilayer, relabeled from Mariana Ruiz Villarreal, public domain
https://commons.wikimedia.org/wiki/File:Bilayer_scheme.svg

Hydrogen bond, relabeled from public domain
https://commons.wikimedia.org/wiki/File:Hydrogen-bonding-in-water-2D.png

Sodium ion solvation, incorporated from public domain
https://commons.wikimedia.org/wiki/File:Na%2BH2O.svg

Messengers and Receptors

Ligand binding, relabeled from public domain
https://commons.wikimedia.org/wiki/File:The_External_Reactions_and_the_Internal_Reactions.jpg

Active site, relabeled from Tim Vickers, public domain
https://commons.wikimedia.org/wiki/File:Induced_fit_diagram.svg

Ligand affinity, modified from Keith Larence Brain, public domain
https://commons.wikimedia.org/wiki/File:Agonists_v2.png

Autonomic Nervous System

Neuron, relabeled from public domain
https://www.wpclipart.com/medical/anatomy/nervous_system/neuron/neuron.png.html

Autonomic nervous system, relabeled from public domain
https://commons.wikimedia.org/wiki/File:The_Autonomic_Nervous_System.jpg

ANS Neurotransmitters

Synapse, relabeled from public domain
https://commons.wikimedia.org/wiki/File:Synapse_blank.png

Phospholipid bilayer, modified from Mariana Ruiz Villarreal, public domain
https://commons.wikimedia.org/wiki/File:Bilayer_scheme.svg

Adrenergic receptors, relabeled and modified from pubic domain
https://commons.wikimedia.org/wiki/File:The_External_Reactions_and_the_Internal_Reactions.jpg

Membrane Transport

Diffusion in solution, J Krieger, public domain
https://commons.wikimedia.org/wiki/File:Diffusion_(1).png

Membrane diffusion, Mariana Ruiz Villarreal, public domain
https://commons.wikimedia.org/wiki/File:Scheme_simple_diffusion_in_cell_membrane-en.svg

Facilitated diffusion, Mariana Ruiz Villarreal, public domain
https://commons.wikimedia.org/wiki/File:Scheme_facilitated_diffusion_in_cell_membrane-en.svg

Facilitated diffusion, modified from Mariana Ruiz Villarreal, public domain
https://commons.wikimedia.org/wiki/File:Scheme_simple_diffusion_in_cell_membrane-en.svg

Active transport, relabeled from Mariana Ruiz Villarreal, public domain
https://commons.wikimedia.org/wiki/File:Scheme_sodium-potassium_pump-en.svg

Neurons

Phospholipid bilayer, modified from Mariana Ruiz Villarreal, public domain
https://commons.wikimedia.org/wiki/File:Bilayer_scheme.svg

Skeletal Muscle Fibers

Skeletal muscle, relabeled from SEER training module (NCI), public domain
https://commons.wikimedia.org/wiki/File:Skeletal_muscle.png

Muscle metabolism relabeled from public domain
https://commons.wikimedia.org/wiki/File:Graphique_%C3%A9nerg
%C3%A9tique.svg

Skeletal muscle fiber, relabeled from SEER training module, public domain
https://commons.wikimedia.org/wiki/File:Skeletal_muscle_and_fiber.jpg

Length-tension curve, relabeled from public domain
https://commons.wikimedia.org/wiki/File:Lengthtension.jpg

Cardiac Physiology

Heart and lungs, relabeled from Gray's Anatomy, 1918, public domain
https://commons.wikimedia.org/wiki/File:Heart-and-lungs.jpg

Heart wall, relabeled from NIH NLM, public domain
https://ghr.nlm.nih.gov

Flow of blood through heart, relabeled from NIH NLM, public domain
https://ghr.nlm.nih.gov

Heart valves, relabeled from Gray's Anatomy, 1918, public domain
https://en.wikipedia.org/wiki/File:Gray495.png

Heart section, relabeled from Gray's Anatomy, 1918, public domain
https://commons.wikimedia.org/wiki/File:Gray498.png

Systole and diastole, relabeled from Mariana Ruiz Villarreal, public domain
https://commons.wikimedia.org/wiki/File:Bluebaby_syndrom.svg

Cardiac conduction system, relabeled from NIH NLM, public domain
https://ghr.nlm.nih.gov

Cardiac and Smooth Muscle

Adherens junctions, relabeled from Mariana Ruiz Villarreal, public domain
https://commons.wikimedia.org/wiki/File:Cell_junction_simplified_en.svg

Blood Pressure

Aorta, modified from Gray, *Anatomy of the Human Body,* public domain
Wikimedia Commons File:Gray506.svg

Tissue Perfusion

Plaque accumulation, relabeled from CDC, public domain
https://commons.wikimedia.org/wiki/File:Coronary_heart_disease.PNG

Lymph capillaries, relabeled from SEER (NCI), public domain
https://commons.wikimedia.org/wiki/File:Illu_lymph_capillary.png

Ventilation

Respiratory tract, relabeled from NCI, public domain
https://commons.wikimedia.org/wiki/File:Illu_conducting_passages.svg

Upper respiratory tract, SEER Training Module (NCI), public domain
https://commons.wikimedia.org/wiki/File:Illu01_head_neck.jpg

Bronchi, relabeled from NCI, public domain
https://commons.wikimedia.org/wiki/File:Illu_larynx.jpg

Pleurisy and pneumothorax, relabeled from NHLBI, public domain
https://commons.wikimedia.org/wiki/File:Pleurisy_and_pneumothorax.jpg

Inspiration and expiration, relabeled from Mariana Villarreal, public domain
https://commons.wikimedia.org/wiki/File:Inhalation_diagram.svg and
https://commons.wikimedia.org/wiki/File:Expiration_diagram.svg

Lung volumes, relabeled, public domain
https://commons.wikimedia.org/wiki/File:LungVolume.jpg

Renal Anatomy and Filtration

Urinary system, relabeled from SEER Training Module, public domain
Wikimedia Commons File:Illu_urinary_system_neutral.png

Kidney, NCI, public domain
https://commons.wikimedia.org/wiki/File:Kidney_and_adrenal_gland.jpg

Nephron, relabeled from Burton Radons, public domain
https://commons.wikimedia.org/wiki/File:Nephron_illustration.svg

Glomerulus, relabeled from Kelly et al., *Diseases of the Kidney* (1922), public
domain
https://commons.wikimedia.org/wiki/File:Diseases_of_the_kidneys,_ureters_and_blad
der,_with_special_reference_to_the_diseases_of_women_(1922)_(14579191128).jpg

Glomerulus, relabeled from Gray's *Anatomy* (1918), public domain
https://commons.wikimedia.org/wiki/File:Gray1130.svg

Cerebrovascular autoregulation, relabeled from public domain
https://commons.wikimedia.org/wiki/File:Cerebrovascular_autoregulation.svg

Digestive Tract

Digestive system, relabeled from Sobotta's Atlas (1906), public domain
https://en.wikipedia.org/wiki/File:Sobo_1906_323.png

Anatomy of abdomen, relabeled, public domain
https://commons.wikimedia.org/wiki/File:Anatomy_Abdomen_Tiesworks.jpg

Small intestine, relabeled from Davison, *Human Body* (1908), public domain
https://commons.wikimedia.org/wiki/File:The_human_body_and_health_-_an_
elementary_text-book_of_essential_anatomy,_applied_physiology_and_practical_
hygiene_for_schools_(1908)_(14749934976).jpg

Intestinal villus, relabeled, public domain
https://commons.wikimedia.org/wiki/File:Intestinal_villus_simplified.svg

Digestion and Absorption

Digestive system, relabeled from Sobotta's Atlas (1906), public domain
https://en.wikipedia.org/wiki/File:Sobo_1906_323.png

Sensory Physiology

Two-point discrimination, relabled from House & Pansky, *A Functional Approach to Neuroanatomy* (1960) public domain
https://commons.wikimedia.org/wiki/File:Lawrence_1960_8.11.png

Lateral inhibition, relabled from Tom Sulcer, public domain
https://commons.wikimedia.org/wiki/File:Lawrence_1960_8.11.png

Meissner's corpuscle, relabeled from Gray's *Anatomy* (1918), public domain
https://commons.wikimedia.org/wiki/File:Gray936.png

Pacinian corpuscle, relabeled from Gray's *Anatomy* (1918), public domain
https://commons.wikimedia.org/wiki/File:Gray936.png

Ruffini corpuscle, relabeled from original slide sent by Angelo Ruffini to Sherrington in 1898, public domain
https://commons.wikimedia.org/wiki/File:Ruffini_Corpuscle_by_Angelo_Ruffini.jpg

Spinal cord, modified from Gray's *Anatomy* (1918), public domain
https://commons.wikimedia.org/wiki/File:Gray770-en.svg

Epidural blood patch, public domain
https://commons.wikimedia.org/wiki/File:Epidural_blood_patch.svg

Golgi tendon organ, relabeled, public domain
https://commons.wikimedia.org/wiki/File:Tendon_organ_model.jpg

Muscle spindle, relabeled, public domain
https://commons.wikimedia.org/wiki/File:Muscle_spindle_model.jpg

Spinal reflex from Pearson Scott Foresman, public domain
https://commons.wikimedia.org/wiki/File:Afferent_(PSF).png

Patellar reflex, relabeled, public domain
https://commons.wikimedia.org/wiki/File:Patelar.JPG

Consciousness and Pain

Choroid plexus, relabeled from Gray's *Anatomy* (1918), public domain
https://commons.wikimedia.org/wiki/File:Gray768.png

Tracts and fibers, relabeled from Gray's *Anatomy* (1918), public domain
https://commons.wikimedia.org/wiki/File:Gray751.png

EEG Sleep stages, relabeled, public domain
https://commons.wikimedia.org/wiki/File:Sleep_EEG_Stage_1.jpg
https://commons.wikimedia.org/wiki/File:Sleep_EEG_Stage_4.jpg

Reticular formation, relabeled from Gray's *Anatomy* (1918), public domain
https://commons.wikimedia.org/wiki/File:Gray690.png

Somatic Motor and Executive Functions

Limbic system, National Institute of Drug Abuse for Teachers, public domain
https://commons.wikimedia.org/wiki/File:Brain_limbicsystem.jpg

Cerebral lobes, relabeled, public domain
https://commons.wikimedia.org/wiki/File:Brainlobes.png

LTP, relabeled, Tom Sulcer, public domain
https://en.wikipedia.org/wiki/File:LTP_First_Stage.png
https://commons.wikimedia.org/wiki/File:LTP_Fourth_Stage.png

Memory consolidation, relabeled, public domain
https://commons.wikimedia.org/wiki/File:Ribotlawstandardmethoddiagram.jpg

Dopamine, relabeled, National Institute of Drug Abuse, public domain
https://en.wikipedia.org/wiki/File:Pubmed_equitativa_hormonal.png

Basal ganglia, NIH, public domain
https://commons.wikimedia.org/wiki/File:Basal_Ganglia_lateral.svg

Speech centers, NIH, public domain
https://commons.wikimedia.org/wiki/File:BrocasAreaSmall_(ja).png

www.ingramcontent.com/pod-product-compliance
Lightning Source LLC
Chambersburg PA
CBHW071251220526
45468CB00001B/74